THE NEW NATURALIST
A SURVEY OF BRITISH NATURAL HISTORY

FARMING AND WILDLIFE

EDITORS
Margaret Davies, C.B.E., M.A., Ph.D.
Kenneth Mellanby, C.B.E., Sc.D.
S. M. Walters, M.A., Ph.D.

PHOTOGRAPHIC EDITOR
Eric Hosking, F.R.P.S.

The aim of this series is to interest the general reader in the wild life
of Britain by recapturing the inquiring spirit of the old naturalists.
The Editors believe that the natural pride of the British public in the
native fauna and flora, to which must be added concern for their
conservation, is best fostered by maintaining a high standard of
accuracy combined with clarity of exposition in presenting the
results of modern scientific research.

THE NEW NATURALIST

FARMING AND WILDLIFE

KENNETH MELLANBY, C.B.E., Sc.D.

With 67 photographs in black and white

COLLINS
ST JAMES'S PLACE, LONDON

William Collins Sons & Co Ltd
London · Glasgow · Sydney · Auckland
Toronto · Johannesburg

First published 1981
© Kenneth Mellanby 1981
ISBN 0 00 219239 X
Filmset by Jolly & Barber Ltd, Rugby
Black and white reproduction by Adroit Photo-Litho Ltd, Birmingham
Made and printed in Great Britain by
William Collins Sons & Co Ltd, Glasgow

CONTENTS

PHOTOGRAPHS

AUTHOR'S PREFACE

THE aim of the New Naturalist series is 'to interest the general reader in the wildlife of Britain'. As over 80 per cent of the surface of this country is farmed in one way or another, the majority of New Naturalist books have been at least partly concerned with the effects of agriculture on our wild animals and plants. This is particularly true of the late Sir Dudley Stamp's *Man and the Land*, the late Sir John Russell's *The World of the Soil*, Dr Ian Moore's *Grass and Grasslands*, my own *Pesticides and Pollution*, the late Dr R. K. Murton's *Man and Birds* and, perhaps most significantly, *Hedges* by my colleagues Drs E. Pollard, M. D. Hooper and N. W. Moore.

As the series already covers the subject of farming and wildlife so substantially, readers may wonder why this further volume is thought necessary. I believe that it may be valuable to consider, in one volume, the ways in which modern farming is changing our countryside and the wildlife of that countryside. I have tried throughout to concentrate on the effects of agriculture as it is practised in Britain in the second half of the twentieth century.

I could not have attempted to write this book had I not been so fortunate as to have had unique opportunities to see, at first hand, the problems of both farmers and conservationists. I was for over six years head of the Entomology Department at Rothamsted Experimental Station, the world's senior agricultural research station. Then, in 1961, I joined the staff of the Nature Conservancy, on appointment as the first Director of Monks Wood Experimental Station, where the emphasis was on conservation, but where the effects of pesticides and changing farming practices on wildlife were investigated. I have thus had the opportunity of working with colleagues who have been closely involved with my subject. Thus lowland grassland was the main interest of the team led by Dr Eric Duffey; he has edited an authoritative book *Grassland Ecology and Wildlife Management* which has proved a valuable source of information on this subject. Dr Norman Moore was head of the Pesticides and Wildlife Section, and he also organised the Nature Conservancy's Agricultural Habitat Team, which contributed substantially to my subject. Then I have lived for twenty years in the most intensely farmed arable area of Britain, and have had the opportunity of discussing their problems with farmers and land owners throughout the country. So I hope that I have been able to understand the point of view of both farmers and conservationists.

It would be idle to deny that there has been considerable conflict between the two groups. As I show, most recent developments in farming have been harmful to our wild plants and animals. Rare species have become rarer or extinct, and even common varieties may no longer be present for the enjoyment of the public in many parts of the country. On the other hand many conservationists have shown little understanding of the real difficulties of farmers, and have given them no help in overcoming these problems. Fortunately there is now a growing movement to bring the two sides together, to which I hope that this book will make some contribution.

In common with other books in this series, I have used common English names for plants and animals where these are generally accepted. I have also given the Latin name on the first occasion on which a species is mentioned. I have tried always to use the most generally accepted Latin names, relying for instance on *The Flora of the British Isles* by W. R. Clapham, T. G. Tutin and E. F. Warburg for flowering plants, and for insects I have used *A Check List of British Insects* and other publications by G. S. Kloet and W. D. Hincks. I apologise to any systematists if I have not given any new Latin names which their studies have established; I think that readers will be able to identify the organisms which I mention in my text.

CHAPTER I

INTRODUCTION

GREAT BRITAIN became an island, separated from continental Europe, some seven thousand years ago, when the level of the North Sea rose and covered what was until then dry land connecting the eastern part of England with Denmark and the Low Countries. At this time almost the whole of the British lowlands was covered with deciduous forest, mainly oak. In the mountainous area conifers took the place of deciduous trees, though the highest mountains were open moorland. Grassland only occurred in small temporary patches where the forest had been destroyed or where trees had died of old age. There were substantial areas of bogs and marshes, and the climate was warmer and wetter than it is today. The flora and fauna consisted mainly of woodland species and those living in marshes and on the sea shore.

At this time Mesolithic man was still a hunter and gatherer making little impact on his environment. His numbers were low, perhaps only a few thousand individuals in the whole country. Then about five thousand years ago Neolithic man invaded Britain, bringing pottery, polished stone implements and arable agriculture. This was when man's influence on the countryside first began to be noticeable. Small fields cleared from forest and scrub were cultivated and corn was grown, but livestock, sheep, goats and cattle, were grazed and extended the areas of grass on the chalk and limestone areas such as Salisbury Plain, the South Downs, the Yorkshire Wolds and the Cotswolds. Man's numbers grew as his food supply increased, and farming methods became more sophisticated with the successive invasions of Bronze Age (2,000 B.C. to 500 B.C.) and of Iron Age people (500 B.C. to 40 A.D.). By the time of the Roman occupation of 43 A.D. there were probably half a million Britons, and farming in the southern part of the country was so successful that wheat was exported to what is now France. Woodlands still covered most of the country, marshes and fens were undrained, but there were substantial areas of open fields with cereal crops or grass for grazing livestock.

During the period of the Roman occupation and the unsettled centuries between the withdrawal of the legions in 383 A.D. and the

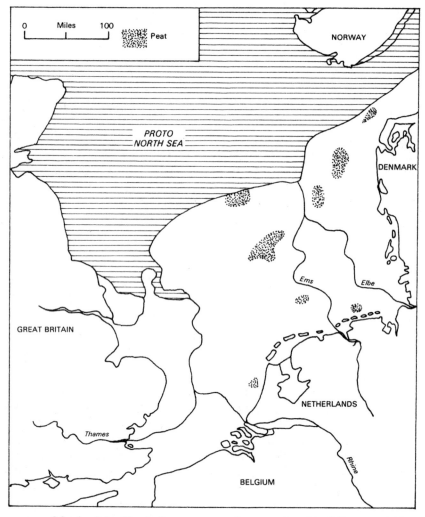

FIG. 1. Much of the North Sea area was above sea level in BC 7,500. (After Stamp 1969)

Norman conquest, farming continued to develop. The Saxons are generally credited with the introduction of the heavy ox-drawn plough which enabled heavy clay soils to be brought into cultivation. More of the forest was removed, and the first hedges grew up to divide fields and parishes. Grazing cattle, by preventing woodland regeneration when they ate tree seedlings or trampled them with their hooves, contributed to the spread of grasslands and to their own food supply. The human population increased slowly with the improving agriculture, reaching about one million by 1000 A.D.

After the Norman invasion in 1066 A.D. until the Black Death in 1348–9 the population of Britain grew more rapidly from one to nearly four million, and farming progressed so as to feed the increasing numbers. Although the Norman kings, who were keen hunters, sequestered substantial areas of 'forest' (much of which was heath and rough ground) where deer had priority, the arable area was probably, by the middle of the fourteenth century, almost as extensive as that which exists today, i.e. nearly 4,000,000 ha. We had the open field system, which in some parts of Britain must have resembled the present day prairie-like landscape of East Anglia, though small enclosed fields were common in Wales. This huge area of arable only barely fed the population, for yields were low, reaching perhaps 600 kg of wheat per ha compared with eight to ten times that amount obtained today. The situation was even worse than these figures suggest, for a fifth of the yield had to be saved for seed corn, and much of the land was left fallow every second or third year to try to restore its meagre fertility. Although the value of manure was at least partly understood, there was little to spread on the arable fields as livestock fed on the grass and returned their excrement to it.

The Black Death wiped out whole villages, and the population of Britain may have been halved. Many arable fields went out of cultivation, and were invaded by scrub and trees. The area under grass was extended and supported a growing population of sheep which may have reached as many as twelve millions in the fifteenth century, and brought great prosperity to a minority in the appropriate districts. This left a legacy in the form of the magnificent 'wool' churches of East Anglia and the Cotswolds. There was sometimes a food shortage even with the small population, which did not again reach the level of that before the Black Death until 1600 A.D., as grass fields for sheep were sometimes preferred to wheat fields for human food. However, the sheep made their contribution, not only from wool and, to a lesser extent, meat, but also by improving the fertility of the land on which they grazed.

In the three hundred years up to 1939 we had a farming revolution which completely altered the appearance of much of Britain, and which increased the production of food many fold. Up till then, winter keep for livestock was scarce, and so only a minimum number of cattle could be retained. New crops like turnips and new rotations produced much more food, and made it possible to support more livestock. This in turn produced more manure, which increased yields particularly of cereals. To make this new system work the land had to be divided up into convenient plots by stockproof hedges or fences. This produced the traditional landscape which most people think of as having existed from time immemorial. In fact it has existed for only a comparatively short time, but it has produced a beautiful landscape and one in which much of our wildlife flourishes.

I am concerned mostly in this book with recent changes in farming practice which have taken place since 1945, and with the effect of these changes on our native flora and fauna. I deal with the specific effects of arable farming, grassland management, animal husbandry, the soil and its inhabitants, hedge removal, drainage and farm chemicals in succeeding chapters. Figure 2 shows how the total area of Britain is being used today. The tiny proportion actually devoted to nature conservation indicates the importance of the much larger area used by all forms of agriculture.

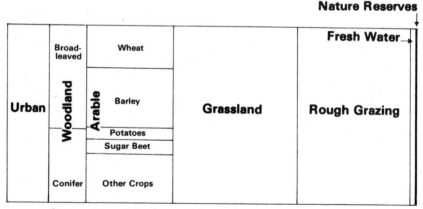

FIG. 2. Comparison of the areas of different land use, and different crops, in Britain in 1980.

What is not always realised is that our landscape, even the 'traditional' landscape which existed before 1945, was almost entirely man-made. Even the open moorlands of Exmoor, which conservationists are at such pains to preserve, or the heather-covered hills of Northern

England and Scotland, which are said to be 'threatened with afforestation', are mostly man-made impoverished communities which were previously covered with trees. Our Bronze and Iron Age ancestors were not good farmers. They cleared the trees, took a few crops which impoverished the soil so that no more could be grown, and then moved on. In some cases grassland which could be maintained with regular grazing by sheep resulted, but in others the end result was very unproductive moorland. It can be said that in many cases the most natural-looking areas in Britain are those where man has made the most drastic transformation.

Britain's native wildlife is generally considered to be that which existed when the land bridge with continental Europe was submerged seven thousand years ago. Not all species that existed at that date still persist, and the relative abundance of most plants and animals has greatly changed. We have already noted that our wildlife was, originally, mostly that of woodland, or of the woodland edge. Many species can only continue to exist in woodland, or in hedgerows which provide some of the conditions of the woodland edge. Others have adapted well to the changed landscape and flourish in open fields or even in suburban gardens. Some species find the new conditions so congenial that they increase so greatly as to become pests. So the populations of our plants and animals have changed markedly over the last seven thousand years.

Even before Neolithic man became a farmer the enormous Irish elk and the aggressive auroch or wild bull were decreasing in numbers, and were quickly exterminated. The brown bear, wolf, beaver and wild boar survived into historic times, but were extinct or rare by the end of the mediaeval period. These were killed off by man even before changes in the amount of suitable habitat would have made their survival uncertain.

Changes in bird species have also occurred. We have few reliable records of extinctions before 1800, except perhaps for the capercaillie (*Tetras urogallus*), but since that date have lost, among others, the great auk (*Pinguinus impennis*), the great bustard (*Otis tarda*), the avocet (*Recurvirostra avosetta*), the bittern (*Botaurus stellaria*), the ruff (*Philomachus pugnax*), the osprey (*Pandion haliaetus*), and the white-tailed eagle (*Haliäetus albicilla*). Fortunately most of these (the avocet, bittern, the ruff and the osprey) have re-established themselves in recent years. Some of the extinctions, e.g. the great auk, were caused by hunting, some were hastened by egg collectors, but the main cause of the disappearance of the others was loss of habitat. The main area to be lost in the nineteenth century was wetland, when the fens and other marshes particularly in East Anglia which had survived the improvements of the 17th century were themselves drained.

Other losses were among insects. Here we have few reliable records for most species, but know that the beautiful large copper butterfly (*Lycaena dispar*) was last seen in 1851; fen drainage diminished the extent of its food plant, the great water dock, and collecting by entomologists probably completed the process. Other butterflies which we have lost are the black-veined white (*Apocia crataegi*) and, more recently, the large blue (*Maculinea arion*).

The actual extermination of a species often causes the most concern, even when the creature has been so rare for many years that few have seen it in Britain. Scientists are divided in their reaction to these exterminations. If the whole world population is destroyed, so that an animal or plant becomes extinct, everyone is concerned at this impoverishment of the globe. However, many of the British rarities are common in other parts of Europe; Britain may be at the extreme limit of their natural range, and any small change in climate could make this country less suitable. Perhaps we should be more concerned with serious reductions in the populations of commoner species, which mean that most people no longer have the opportunity of seeing them. This point is discussed in the succeeding chapters in relation to different habitats.

We have not only lost animals and plants which existed seven thousand years ago, we have also gained quite a number. We tend to react very differently depending on the way in which such introductions took place. Ornithologists and entomologists, dealing with birds and insects, have generally a different point of view from that of botanists, concerned with plants, or even from those concerned with mammals.

The true native wildlife, that is the wildlife which was present in 5000 B.C., is esteemed by conservationists, and every effort, however futile in practice, is made to ensure that these species continue to live in Britain. In general if a species is added to this original list, if it arrives in Britain and establishes itself without human intervention, it is accepted as a welcome addition to the fauna or flora. If man is responsible for the introduction, it may be considered to be an undesirable alien. The more recent the introduction, the more undesirable the organism.

In the case of birds, those which colonise or recolonise sites in Britain are welcomed with open arms. Thus the avocet which became extinct as a breeding species in 1844 nested here once more in 1946; it was then chosen as the emblem of our leading ornithological organisation, the Royal Society for the Protection of Birds. The other species which formerly bred here but became extinct were made equally welcome. In particular the osprey, last breeding in 1903, was hailed as a triumph of conservation when it bred again in the highlands of Scotland in 1954. The bittern, black-tailed godwit, ruff and Savi's warbler have been accepted with enthusiasm.

Many of the birds seen in Britain are migrants which overwinter here but breed in other parts of the globe. One such species is the Bewick's swan, increasingly frequently seen in winter in the Ouse Washes and the Severn estuary, which breeds in Siberia. This is looked upon as a welcome addition to our wildlife, and reserve areas are managed to encourage it to visit them. I think it is true to say that all self-propelled migrants are accepted. The situation is quite complex, for some birds are regularly seen as migrants and only occasionally as breeders. When breeding does take place within our shores, this is generally welcomed.

Thus several water birds first nested here in the nineteenth century, and several others, including the attractive firecrest, have established themselves since 1945. The only possible exception, where our reaction is less enthusiastic, is the collared dove. This species was first recorded breeding in Britain in Norfolk, in 1955 and it spread slowly and in small numbers over East Anglia in the next few years. It was welcomed at first, but then it proved itself to be only too successful. It had clearly found a vacant niche. The species now occurs all over the British Isles, and is common in suburban and rural gardens, where it is something of a pest, attacking vegetables and other crops. Even its penetrating and monotonous song is considered unattractive. So it still bears some of the stigmata of the alien.

During the second half of the nineteenth century there were many 'Acclimatisation Societies' in Britain and other countries. The object of their members was to acclimatise desirable species and introduce them to new countries, where they would, it was believed, be useful and attractive. So a whole series of birds and mammals were taken to Australia and New Zealand, mostly with very unfortunate results. Thus the rabbit (*Oryctolagus cuniculus*) overgrazed the vegetation and the fox (*Vulpes vulpes*) destroyed the smaller native marsupials. Most introductions of birds to Britain failed to establish themselves, but

some, including the little owl (*Athene noctua*), succeeded. This bird is a
native of Southern and Central Europe and much of Asia, and is the
emblem of Pallas Athene (and of the London club, the Athenaeum).
Many well-meaning naturalists released these owls from 1842 on-
wards. Those reading the original accounts may find them confusing.
Thus in a letter to a friend, written in 1889, Lord Lilford writes: 'I
turned down about forty little owls'. This does not mean the birds were
destroyed: 'turned down' is the expression used in sporting circles
when birds are released in order to try to stock an area.

The little owl established itself, and numbers became so great in the
1930s that there were complaints that it was damaging game birds. A
careful study showed that the bird fed mainly on insects, it also ate
small rodents, and that game birds were hardly ever taken. Clearly the
little owl found a vacant niche for a day-flying insectivorous bird. The
species is still widespread, but the increase in numbers did not continue
and it is less common than it was fifty years ago. It is now accepted as a
British bird and is even given legal protection.

Although the Romans have been credited with introducing pheasants (*Phasianus colchicus*), they were probably brought to Britain in the
twelfth century as captive breeding birds, and were prized for eating.
The pheasant became naturalised in the countryside in the fourteenth
or fifteenth century. Several species of pheasant have been introduced,
and hybrids occur. Today the feral pheasant still abounds – on February 9th, 1980 eight, six hens and two cocks, were strutting about on
the lawn just outside the drawing room window of my farm house in
Cambridgeshire. The stock of pheasants is augmented by captive
rearing and release of large numbers in those parts of Britain where
shooting is popular. Many arable farmers who would otherwise prefer
to remove all obstructions to mechanised cultivation are willing to
retain some hedges and areas of cover to provide nesting sites for
pheasants. This bird, the red legged partridge (*Alectoris rufa*) and other
game birds are considered in chapter 11 (p.132).

The Canada goose (*Branta canadensis*) was first introduced as a
captive, pinioned, species, but is now accepted as a 'wild' British
species. Other wildfowl have been similarly introduced, and some
have had wild populations augmented by introduced stock. In some
parts of eastern Britain geese are considered to be pests as they graze
both grass and cereals.

Mammals are unable to reach Britain except when aided by man.
Some, like the rabbit and the grey squirrel (*Sciurrus carolinensis*), were

brought here deliberately and released into the countryside. Others, like the house mouse (*Mus musculus*) and the rats (*Rattus* spp.), came as stowaways on ships. The coypu (*Myocaster coypus*) and the mink (*Mustela vison*) were brought as captives for fur farms and escaped. Fallow (*Dama dama*) and other deer were imported to grace the parks of the nobility; some escaped, others were deliberately liberated. There have been escapes from zoological gardens, and a few species, including the red-necked wallaby (*Macropus rufogriseus*) and the porcupine (*Hystrix* spp.) have been able to establish small British populations.

The rabbit is classed as an agricultural pest, and strenuous efforts are made to keep numbers down or even to eradicate it. This species was introduced in the twelfth century, and it remained in semi-captivity in most parts of the country for some centuries. Although in the fourteenth century there were complaints that crops were being damaged, the rabbit only became common in many areas in the nineteenth century. By the 1930s there may have been as many as 100 million rabbits in Britain, and the damage they did was substantial. Control by myxomatosis in 1953 was at first successful in almost wiping them out in many places, but the species now fluctuates in numbers, the disease usually striking again when the population reaches a damaging level. (See chapter 3.)

The grey squirrel was recorded in Wales in 1828, but we do not know who brought the animals from America. Later, from 1876 to 1929, a great many were 'turned down' in more than twenty sites in England, Scotland and Wales. It was not until 1938 that importation of this pest was made illegal. The grey squirrel is now found in most parts of England and Wales and in smaller areas in Scotland and Ireland. It does considerable damage to orchard, garden and farm crops, and is particularly harmful to trees, preventing natural regeneration in beech woods. It is often blamed for the fall in numbers of the native red squirrel (*Sciurus vulgaris*), but it seems likely that this species was suffering from epidemic disease before its rival appeared.

Both the black (*Rattus rattus*) and brown rats (*Rattus norvegicus*) have been imported accidentally, the first with the returning crusaders in the twelfth century bringing the flea that transmitted the Black Death, the latter on timber-carrying ships some six hundred years later. No one has a good word for either species; unfortunately they found conditions in Britain which proved very favourable, and even the modern farm still supports a substantial population of the brown rat.

The importance of these species in spreading disease is discussed in chapter 11 (see p.145).

Wallabies have escaped from zoos and established small breeding populations in the Peak District of Derbyshire and Staffordshire, and two hundred miles away in Sussex. These have existed for nearly 40 years, and even survived the severe winter of 1962–3. Unlike most escapes these animals have been welcomed by many naturalists. They do no recognisable harm and seem unlikely to spread to very large areas. Another escape from zoos, the porcupine, is found in Devon and Derbyshire. (Two different but related species are involved, *Hystrix hodgsoni* invading Devon, *H. cristata* Derbyshire.) It has not been welcomed, as it is potentially a serious pest, seriously damaging timber plantations. Attempts are being made to eradicate the porcupine, and also other dangerous aliens which have escaped from captivity, including the coypu and the mink.

The list I have given here is incomplete. As many as twenty-seven of the sixty-seven terrestrial mammals found 'wild' in Britain are not truly indigenous. The small numbers of reptiles and amphibia have also been greatly augmented by introductions. When Britain became an island, it had only a few of these cold-blooded vertebrates. The last ice age must have exterminated wildlife from most of Britain, and when the ice retreated north the climate would be, for some centuries, too cold to allow either reptiles or amphibia to live here. There was only a comparatively short period of perhaps two thousand years when it was warm enough to allow migration into Britain before the land-bridge disappeared. As far as I know none of the indigenous amphibians or reptiles has been exterminated, though frog (*Rana temporaria*) numbers have been reduced and the natterjack toad (*Bufo calamita*) is only found in a few of its former sites. Among reptiles the smooth snake (*Coronella austriaca*) and the sand lizard (*Lacerta agilis*) are also considered as endangered species. On the other hand six amphibia, including the edible frog (*Rana esculenta*), are now reasonably if locally established in England.

Insects, like birds, may reach Britain by their own efforts, though aided by winds and air currents. Collectors of butterflies have treasured specimens of the 'black-veined brown' or milkweed (*Danaus plexipus*) which have been blown across the Atlantic from America, and of the Camberwell beauty (*Nymphalis antiopa*), a vagrant which probably emerged from a pupa in Scandinavia. These do not breed when they reach our shores. Other butterflies, such as the painted lady (*Vanessa cardui*) and the red admiral (*V. atalanta*), may breed in Britain, but most of the adults seen have arrived as migrants. Although all these insects are admitted to the 'British list', they do not depend for their survival on a suitable habitat in this country.

Less welcome insects have also invaded this country. The Colorado beetle (*Leptinotarsa decemlineata*) has invaded continental Europe from America, and can devastate potato crops. Fortunately it has so far been eliminated quickly when populations have started to build up in Britain. However, the bed bug (*Cimex lectularius*), which arrived at about the same time as the black rat, still infests some old buildings, though modern insecticides have greatly reduced its incidence.

Many people, when they think of 'wildlife', think only of animals. They forget that were it not for plants there would be no animal life on earth. Plants capture the energy in sunlight and are the only primary sources of food. Plants contribute in many ways to the environment in which animals live. Ecology is the study of animals and plants and the environment in which they live, and conservation of animals without conservation of plants is impossible.

I have already pointed out that the plant cover of Britain has been greatly changed by farmers over the last seven thousand years. Most of the lowlands of southern Britain were covered with deciduous forest in which oak (*Quercus robur* and *Q. petraea*) was dominant; today forests cover less than five per cent of the countryside of England, and much of this consists of exotic conifers. However, none of the indigenous trees has become extinct, though they are all greatly reduced in numbers. Only a comparatively small number of flowering plants have disappeared, though many which were once common are now rare. Loss of habitat to agriculture is the main cause, with drainage of marshes (again to produce farm land) an important factor.

Introduced plants dominate most of the countryside. At least 700 species, no less than one third of the whole listed flora, have been introduced. This process has gone on for a very long time, and trees like the sycamore (*Acer pseudoplatanus*), the horse-chestnut (*Aesculus hippocastanum*) and the larch (*Larix decidua*) are generally accepted as British. The sycamore is sometimes considered a pest, as it spreads rapidly and may crowd out other trees, but it is clearly well adapted to our conditions.

Many introduced plants are now 'wild'; these may be garden escapes, or they may have been deliberately planted in uncultivated ground. However, cultivated exotics cover most of Britain. The crops of arable land, wheat, barley, sugar beet and potatoes to name only the commonest, cover five million ha, and temporary grass leys containing few native plants cover half that area. There is little consolation in the knowledge that most of the important weeds of arable crops, plants like charlock (*Sinapsis arvensis*), blackgrass (*Alopecurus myosuroides*), twitch or couch (*Agropyron repens* and *A. stolonifera*) and mayweed (*Tripleurospermum maritimum*) are indigenous. They must have been rare before arable fields were created. The only major weeds which are imports are the wild oats (*Arvena fatua* and *A. ludoviciana*) which were probably introduced with seed. Actually some of the indigenous weeds have probably been reinforced with new seed in the same way.

Of the five million ha of permanent grass much has been 'improved' and bears mostly non-native species or non-native strains of species which may be considered indigenous. Only in the areas of moorland and rough grazing do native species predominate, for the million hectares of forest consist mainly of introduced trees. British wild plants are thus confined to part of the uplands (and even here the conditions have been radically affected by man), to Nature Reserves (which make up less than 0.5 per cent of our land area), and to patches on farms, in gardens and open spaces where their existence is tolerated.

The situation is even worse than might appear from a look at the map. Nearly a third of the area of Britain – the uplands – would seem to be an area in which wild species might have a good chance of survival. This is the area with the worst soil and the harshest climate, and it remains as it does only because it is generally considered not worth developing. Where marginal land appears to have more potentialities there are pressures to develop it, and these are almost certainly harmful to wildlife. But it is a fallacy to assume that most wild plants, and wildlife generally, are naturally restricted to the least fertile land. The exact opposite is the truth. Even today a glance at any series of distribution maps such as the 'Atlas of the British Flora' produced by the Botanical Society of the British Isles shows that there are far more species of plants in the warmer and fertile parts of Britain than exist in the more barren uplands. Some species which are now restricted to the uplands were, in the past, common in the lowlands. But many lowland species can not survive the harsh conditions of the mountains of Scotland or Wales, so unless they can be preserved in areas of intensive arable farming, they will be eliminated from Britain. Unless we can have nature reserves on high grade farmland we must try to preserve such plants, and many animals also, on our farms.

The purpose of a farm is to produce food. Any organism that seriously reduces food production is a pest, and the farmer is justified in trying to eliminate it. However, many wild plants and animals can exist on farmland without seriously affecting the crops and the live-stock. Individual farmers differ in their reactions to these relatively harmless plants and animals. Some wish to eliminate any which are not of direct value, others enjoy, or at least tolerate, wildlife which does no actual harm. These differing views can both claim scriptural support. Thus in St Matthew's gospel (Chapter XII, verse 30) we find 'He that is not with me is against me'; this text appeals to those who consider that every plant which is not a crop is a weed, and every wild animal is a pest only fit for elimination. On the other hand St Luke writes (Chapter IX, verse 50) 'For he that is not against us is for us'. This, in a farming context, could be taken to suggest that only animals and plants which cause actual damage should be attacked, and that other wildlife should at least be tolerated. St Luke's text is the one with most appeal to the conservationist.

Farmers not only wish to eliminate pests which damage their crops; they also wish to obtain the best yield possible from their land. The good farmer is careful not to try to get a high yield at the expense of damage to his land – he looks for a sustainable return, and if possible adopts a system which maintains and even enhances the fertility of his soil. Unfortunately almost all methods of increasing fertility and im-proving yields, such as improving grassland (chapter 3) or draining swamps (chapter 7) are harmful to wildlife. In chapter 12, I discuss the ways in which conservation can be encouraged with the active co-operation of farmers to the possible mutual advantage of both them and wildlife. But it must be admitted that, if we think only in terms of yield per acre and of maximising the cash value from each hectare of land, then most wildlife has little to offer. There are beneficial insects and birds which reduce pest damage, a healthy soil fauna contributes to fertility, but few conservation measures can be guaranteed to en-hance the farmer's income.

Yet wildlife will not flourish on farmland without the co-operation of the farmer. We do not wish to go back to the Mesolithic landscape, which was mainly dull and, to the forester, rather scruffy woodland with many dead or dying trees. Although wild animals were widely present, many of the birds and plants we now most treasure must have been scarce and difficult to see. Many of the features of the country-side which we most appreciate arise because of the practices of past generations of farmers. The trouble is that modern farming methods have such different side effects. Farmers cannot afford to retain outdated techniques for the benefit of wildlife unless they obtain some tangible benefit in return.

To encourage wildlife is a difficult and complicated process. Plants need somewhere to grow without the wrong type of disturbance. Some flourish best when the area is grazed by livestock at the right time and at the right intensity. Animals which are resident need the correct food, sites to rest and sites to breed. Animals which use farmland for only part of the year may only need resting and feeding facilities. No farm is a self-contained unit, and may provide some facilities which supplement those of, for instance, nearby woods. Streams and ponds plus farmland may support many more wild creatures than do any one of these alone. Some modern farming practices, including the unwise and excessive use of pesticides and fertilisers, may not only damage wildlife within the farms' boundaries, but over a much wider area. This was demonstrated twenty years ago when organochlorine pesticides used on arable fields turned up in birds of prey hundreds of miles away from the places of application. It is also demonstrated by the eutrophication which occurs when fertilisers are washed off farmland and over-enrich lakes and rivers so that blankets of green algae damage other plants and much of the animal life. To get the best results everyone, farmers and conservationists alike, need to understand the long-term effects of all changes in land use throughout the countryside. The following chapters of this book attempt to analyse the ways in which different farming activities, as generally practised today, affect both wildlife and the countryside generally.

CHAPTER 2

ARABLE FARMING

WHEN the first man (or, more probably, woman) made a hole in the ground with a stick or a deer's antler in order to plant a seed, he started a process which has led to profound changes to the surface of the globe and to the plants and animals living on it. As crop growing became more organised, these changes began to be obvious. With the invention of the plough, agriculture as a major industry appeared. The developments of recent years, such as mechanisation and the increased use of fertilisers and pesticides, have simply continued a process which has a long history.

Arable farming (the word is derived from the Latin *arare*, to plough) consists, in essence, of eliminating the natural vegetation from a piece of land and replacing it with a monoculture of some plant which the farmer wishes to grow. Most crops differ from the natural species found in the area where they are grown, and are often different from any plants which grow in the wild. They are, in fact, man-made varieties which have qualities such as high yield or rapid maturation which man finds desirable. Ecologically the result is very unstable and crops can only be produced satisfactorily if they are tended throughout their period of growth. As Sir John Lawes showed more than a hundred years ago at Rothamsted, if a field of wheat is left to itself after harvest, though the shed grain germinates and gives a good cover of wheat plants the next spring, these are quickly overwhelmed by weeds, no crop is obtained even in that year, and no more wheat plants can be found after a couple of seasons. The field is then invaded by scrub followed by trees until something like the landscape of Mesolithic Britain results. In fact, wildlife takes over.

As I showed in the last chapter, crop yields on arable land have greatly increased. The mediaeval farmer obtained under ten bushels of wheat per acre, that is 0.6 tonnes per hectare. The average yield in Britain had risen to 2.85 tonnes per ha in 1952 and to 4.9 tonnes in 1977. At all times the best farmers on good land obtained yields well above the average, reaching today nearly ten tonnes per ha under optimum experimental conditions. However, the low-yielding mediaeval open field was, if anything, less favourable for wildlife than the present-day cereal crop. It may have supported more native weeds, but levels of organic matter were low and the soil fauna was impoverished. A return to mediaeval agriculture would do little to encourage our flora and fauna.

This book is mainly concerned with the ways in which modern farming affects our wildlife, so we must examine modern processes and try to evaluate their effects. As I have just indicated, within the areas devoted to arable crops most wildlife has always been at a disadvantage. It is our elimination of hedges between arable fields, our improvement of ditches draining them and our treatment of marginal land in their vicinity which, in recent years, have had the major, new effects. Nevertheless the way in which our main arable crops – barley, wheat, potatoes, sugar beet – is grown is important both to the farmer and to the naturalist.

In Britain in 1977 4,863,000 ha (12,016,000 acres) were devoted to arable crops. Barley took 2,400,000 ha and yielded ten-and-a-half million tonnes of grain. Wheat covered the second largest area, 1,076,000 ha, and produced more than five million tonnes of grain. Potatoes were grown on 232,000 ha and sugar beet on 201,000 ha. These and other crops grown less extensively (roots, kale, rape, flax, vegatables) took up roughly twenty per cent of the total land area (24,102,000 ha) of the United Kingdom.

In the period since the 1939–45 war the arable area in Britain has decreased by some three hundred thousand ha; this land has been lost to developments of various kinds – housing, factories, roads and motorways. Little of the lost area has developed any great importance for wildlife, though motorway verges may become 'linear nature reserves'. Fortunately productivity on the remaining arable has increased greatly, and the total food production today is not far short of double that obtained thirty years ago. The measures which have ensured this greater productivity have had their effects on wildlife.

In 1939 most fields in Britain were small, many below two hectares. Today, particularly in the main cereal growing areas like East Anglia, most of the fields are very large, often 100 ha or more. Large fields have been created to allow the efficient working of large farm machines. As these get bigger, so do the fields. A few years ago it was thought that an area of 20 ha was the optimum for the tractors and combine harvesters then in use; larger fields simply meant that machines had to make several extra journeys and had to pass through the gates more often.

Since then machines have become even larger, they can operate for longer on a tank of fuel and so they can be used on larger and longer fields.

Enlarging fields has meant the removal of many hedges. The effects of this are described in chapter 6. Quite apart from any loss of wildlife arising from hedgerow removal the survival of wildlife in the crop itself is different in big and small fields. Several animals seem only to venture for a comparatively short distance into an arable crop. With a small field, the animal will be found all over its surface. When the fields are more extensive, only the strip around the edge will be inhabited. Thus the mole (*Talpa europaea*) has its burrow destroyed when a field is ploughed. A few moles are killed, the majority escape to the edge, to hedges or ditch banks where these exist. When cultivation has been completed and the crop sown, the moles make their way back. When travelling by train in early spring it is often possible to look down on fields of winter cereals and to see mole burrows penetrating for fifty or so metres from the edge. Further in, the crop is free from moles and, probably, from some other forms of wildlife.

There are a few mammals and birds which seem well adapted to living in the open areas of the largest field. These include the long-tailed field mouse (*Apodemus sylvaticus*), the brown hare (*Lepus carpensis*), the skylark (*Alauda arvensis*), the lapwing (*Vanellus vanellus*) and the partridge (*Perdrix perdrix*) (though partridges have in fact decreased drastically in numbers for special reasons, see p.133). Increased field size has been deleterious, to a greater or lesser extent, to most other wildlife except for those that are considered to be pests.

A large area of a crop is clearly a wonderful opportunity for any rapidly-breeding insect for which it is the food. Many such insects rapidly increase their populations to pest proportions, and may seriously reduce the yield. Others can do even more serious harm with smaller numbers if they are vectors of viruses which cause diseases in the crop.

It may give some satisfaction to British conservationists to know that most of our insect pests are genuine native wildlife, that is to say they are indigenous species. They were, like many weeds, comparatively uncommon until man provided a habitat which gave so much food that they could increase their population enormously. However, though most pest species were present in Britain over 7,000 years ago, not all the individual insects which attack our crops are of true British provenance. For instance aphids may be carried for several hundred kilometres by air currents at high altitudes. Many perish in the ocean, but British insects reinforce continental populations and vice versa.

Pest insects may be controlled by chemical insecticides, which may themselves pose problems for wildlife, a problem discussed in chapter 9. Field size itself does not seem to be correlated either way with damage by pests and diseases. It has been suggested that crops grown in large blocks are at greater risk than those grown in small ones. In theory one might expect that the small field would sometimes be missed by migrating pests, and that it would also allow for better control by beneficial predators which could migrate in from the peripheral areas. On a garden scale a single row of beans may be saved from serious damage by aphids if it is surrounded by patches of nettles in which predators may overwinter. Experience suggests that even the smallest field which is an economic proposition is too large for this sort of control to be effective. There is even some evidence that pests, which enter a field at its edge, may, like some other forms of wildlife, be least numerous in the middle of enormous fields.

Weeds of arable crops reduce yields and make mechanical harvesting difficult. Today improved seed cleaning means that fewer weeds are planted with the crop, but some weed species survive for many years in the soil to germinate many years later. Other seeds are brought in by the wind and by birds. Herbicides are making modern crops less seriously infested with most weeds, as described in chapter 9. As weeds are mostly British, this can be said to be reducing wildlife, and some so-called weeds (e.g. field pansy, *Viola arvensis*), probably do little harm to crops and so could well be tolerated. Some weeds, like charlock (*Sinapsis arvensis*), seem to have adapted to living in arable crops, which can, in spring be almost smothered in their yellow flowers (if herbicides are not used to control them) but have become virtually extinct as true wild plants. Some years ago I left a field which had grown arable crops for hundreds of years to regenerate naturally. The first year it was so well covered with charlock that it looked as if it were being grown as a crop. The plants grew 60 cm high or even higher. The next year grasses and other species spread and there were only the odd, somewhat smaller, plants of charlock. By the fourth year charlock had apparently disappeared, though careful examination showed occasional tiny plants only two or three cm high bearing, in some cases, a single normal flower. The next year not even these dwarfs could be found. The weed was unable to compete with the rest of the wild vegetation, so charlock could be said to have become a plant, though an unwelcome one, of cultivated land.

Most arable crops give increased yields if they receive large dressings of chemical fertilisers. These do little damage to wildlife in the fields to which they are applied (but see p.37 below on organic farming), but they may be leached out and cause eutrophication in drainage ditches, streams, ponds, rivers, lakes and reservoirs. This is not a new process, but has got more serious as fields are better drained and as larger amounts of nitrogenous fertilisers are applied. Even farmyard manure, particularly if it is applied to bare soil in autumn and winter, loses much of its nutrients, but in cold weather these tend to be carried to the sea in flood water without stimulating algae and other plants to excessive growth. Today more and more chemical fertilisers are applied in spring and early summer, and most of this may be taken up by the growing crop. However when we get heavy rainfall on saturated ground the loss is considerable, and the whole balance of life in the enriched water may be upset. Many attractive water plants may be harmed by competition from algae, and when this algal growth dies and rots it may deoxygenate the water and kill both fish and invertebrates. This subject is dealt with in more detail in chapter 7.

In the past after harvest cereal fields provided excellent feeding grounds for many forms of wildlife. Thus the woodpigeon (*Columba palumbus*) bred most successfully in early autumn, because this was the time when its food was most plentiful. Shed grain on the unploughed stubble contributed substantially to that food. Many other species also benefited from this easy supply. After harvest, the majority of fields were left uncultivated for weeks and even months. It was also common for cereals to be undersown with clover, so that in the next year it provided good grazing for livestock. It also supplied food for pigeons during March, the time of greatest food shortage, and allowed many birds which would otherwise have died of starvation to survive the winter.

Today the combine harvester copes with most crops at breakneck speed. The amount of grain dropped on the ground, is considerable, and is often more than was left by the old reaper and binder. This shed grain is a potential food for wildlife. The straw is left on the ground. Much of this, particularly from barley which has a good feeding value for livestock, is baled and removed, but very large amounts of wheat straw, amounting to over three million tonnes in a year, are burned on the ground. Many people are very concerned. The burning causes some air pollution – new wet paint outside on my own house has been ruined with smuts from a nearby field burned, improperly, in a high wind. Burning straw is considered to be a waste of a valuable resource, and also to endanger wildlife. Yet many farmers consider that it is something which they must do in the interests of agricultural efficiency.

The National Farmers' Union (N.F.U.) is aware of the public concern, and of the possible danger to wildlife, and has issued 'The Straw Burning Code'. Where this is strictly observed, environmental damage is reduced to a minimum. The Code starts with an apologia: 'Straw and stubble burning is an important aid to arable cultivation and the more efficiently the operation is carried out the greater the benefits without any increase in the risks involved. Farmers therefore have a vested interest in ensuring that the operation is carried out efficiently, effectively and safely. Farmers also have a special responsibility to preserve the countryside, the landscape AND ITS WILD-LIFE. They must ensure that they do not cause nuisance and danger from smoke and smuts to neighbouring properties, users of the highway and members of the public. Heavy claims have been made against farmers as a result of fires that have been started by sparks and from damage arising when burning has got out of control. Make sure that you are not faced with such claims – that means TAKE CARE!.

The code spells out the details. One of the most important insists that a firebreak at least 30 feet wide is made around any area to be burned by removing the straw from the perimeter of the field and ploughing or cultivating this strip to a minimum width of nine furrows. Farmers are reminded that it is an offence to start fires within 50 feet of a highway. They are advised to listen to weather forecasts warning of high winds to come, when burning must not be done. Burning should be done early in the day, and fires should be out before nightfall. Various provisions to safeguard thatched buildings, hay or straw stacks, standing crops, woodlands, hedgerows, trees and 'any wildlife habitat' are suggested.

The code is to be welcomed. In the past fires have destroyed many hedges and hedgerow trees. It is sometimes suggested that this was done deliberately by tenant farmers whose landlords would not allow them to remove these hindrances to easy cultivation. Unfortunately not all farmers obey the code, but the N.F.U. is using its influence to get 100 per cent observance.

Careful straw burning does not, directly, do much harm to wildlife.

It will undoubtedly kill many insects and other animals on the surface of the field. Some small mammals which are searching for food will also be incinerated but no serious effects on any animal populations have been reported. Weed seeds are said to be reduced, though many of these, and of the grain from the crop, survive. Though the surface of the ground reaches a temperature lethal to life the heat does not penetrate beneath the surface and the soil fauna seems to be little affected.

Farmers obtain several benefits from straw burning. It is cheap and quick. To bale and cart straw generally costs more than the product is worth, at least to purely arable farmers who have no livestock needing straw for bedding or fodder. I have seen fields where dry straw is burned within hours of combining, and the field may then be ploughed without delay. This saving of time can be invaluable in uncertain weather, where the job can be completed before the rain makes the soil unworkable possibly for several months so that winter cereals cannot be planted. The burnt straw adds an immediate dose of nutrient salts, particularly potash, to the soil, and this can contribute to the growth of grain sown in early autumn. Critics of burning say it would be better ploughed in to increase the level of organic matter, and ultimately of humus, in the soil. This may be true as a long term policy, but loose straw on the ground is difficult to bury unless chopped (a further expensive process at a time when farm labour is stretched harvesting other fields), it clogs the machinery, and in its first year it is slow to break down and necessitates the use of an additional dressing of nitrogen if crop yields are to be maintained.

If the field is ploughed immediately after burning, shed grain is buried and is not easily available to pigeons and other birds. However if it germinates it may contaminate a future crop. For this reason the land may be left for about two weeks, in which time the seed germinates. During this period it is available to wildlife, and large numbers of pigeons, starlings (*Sternus vulgaris*), and other species are generally seen on the stubble. The field is then sprayed with a herbicide, usually paraquat, which kills the sprouting grain and any weeds which have germinated. Applied in this way, paraquat (which is very poisonous when wrongly used) has little harmful effect on wildlife, even on the soil fauna. It is quickly adsorbed (i.e strongly bound) onto the surface of soil particles and vegetation, so that it is permanently immobilised. Fields are cultivated and sown very soon after the application of the herbicide. If the ground is ploughed the soil fauna is affected, many species including earthworms being killed. Increasing use is now being made of minimum cultivation, where the seed is sown directly into the bare soil, or when only a very shallow layer is broken up with a rotavator. Under these circumstances soil animals flourish much better than when the ground is ploughed.

I have already indicated that no arable crop provides ideal conditions for wildlife. However, some are better than others. The greatest area is under cereals. Where these are sown in autumn (winter wheat or barley) they provide grazing for geese, rabbits and deer in winter and early spring, times when grass and other food may be scarce. This grazing by farm livestock has often been practised, not only to feed the animals but also to encourage more shoots (tillering) and so to produce a thicker stand of the corn. Spring sown cereals obviously provide no winter food, and are seldom grazed seriously as they come up when other plants are growing actively. Ground-nesting birds make their nests successfully in the cover provided by both winter and spring sown grains. In summer and until harvest many animals shelter in the growing corn, as is obvious when rabbits bolt as the combine harvester cuts the last patch in the middle of the field. The ripening grain is attacked by many seed-eating birds, sometimes to such an extent that economic losses are sustained.

Rape and beans are insect-pollinated, as is clover which may be grown as a crop for seed as well as forming part of a ley which is grazed or cut for hay or silage. The honey produced is considerable, and is welcomed by bee keepers. It must also be valuable for wild bees, which are efficient pollinators, and to many other insects.

Sugar beet and potatoes are planted on bare soil in spring. These crops are comparatively new to Britain, unlike cereals which, in some form, were grown by our first farmers. Neither of these crops is eaten by many wild animals. They both provide some cover during summer and early autumn, the potatoes until the haulms are destroyed by chemical desiccants before harvest, and the beet until it is lifted and taken to the factory to extract the sugar. Beet tops are valuable as cattle fodder, and it is therefore surprising that they are not grazed more by wildlife. Some beet is not lifted until early January, so it will be available for cover for much of the winter. Game birds are often found sheltering among the sugar beet.

Kale, grown to feed livestock, may stand the whole of the winter to be eaten in March or even April when, in a late year, little grass has grown. In the barest parts of Eastern England, where intensive arable farming has left little cover, quarter-hectare patches of kale are grown as cover for pheasants by farmers who are keen on shooting. This kale will also provide a habitat for other wildlife in the bleak countryside.

This short review of our main arable crops suggests that recent changes in farming practice have, in total, made arable crops, which have never played a major part in wildlife conservation, somewhat less valuable to wildlife. The greatest dangers arise from chemicals used on arable land which may find their way into other habitats and damage wildlife there.

Some ecologists have expressed concern about the whole system of modern intensive arable farming, with larger and larger areas of monoculture and, in some cases, the same crop growing on the same ground for many years with no break and no series of rotations. They fear that there will be soil deterioration, erosion and a loss of fertility and a reduction in food production. They suggest that such a catastrophy might be avoided by going back to traditional rotations and to less intensive methods, with a possible gain in better wildlife conservation. From the point of view of the conservationists there is, unhappily, little evidence to support this view. Good farmers, using the most modern techniques such as minimum cultivation (in which the soil is not ploughed and only a shallow surface layer is broken up), are enhancing the fertility of their soil with each succeeding crop. They are even increasing the level of organic matter in the surface layers of the soil without adding any organic matter in the form of manure. The virtue of minimum cultivation is that it reduces the breakdown by oxidation of the soil organic matter which takes place when the soil is turned over by the plough. The roots of cereal crops, which are extensive, produce this increasing amount of soil organic matter. The situation is not as simple as has just been implied. On many heavy soils things usually go as has been indicated. On more difficult soils it may be necessary to take particular care over drainage, and to introduce a grass break every few years. The important point is that, for the first time in our history, good farmers can (weather permitting) continue to obtain high yields, to maintain and even to enhance the fertility of their soil using methods which are, unfortunately, the least favourable for wildlife. This colours the whole argument about farming and conservation which is the topic of chapter 12.

So far I have been considering the effects of orthodox farming, as practised by the majority over most of Britain. However, we have a growing number of 'organic' farmers who work under different principles. They endeavour to use no synthetic fertilisers or agricutlural chemicals, to produce food of higher quality, to avoid polluting the environment, and to work in with wildlife rather than to its detriment. This whole question of different farming methods is a controversial one unsuitable for detailed consideration in this book, but I think it deserves to be included so far as wildlife is concerned. Although our Ministry of Agriculture, Fisheries and Food and the Royal Commission on Environmental Pollution consider that organic farming 'will not become a large part of the agricultural scene' the movement is growing, and some at least of its principles may be more widely adopted particularly as energy costs, and the price of agricultural chemicals, rise.

It is difficult to evaluate organic farming, mainly because the scientific establishment has refused to give it adequate study. I was for some years involved with the long-term experiment of the Soil Association at New Bells Farm in Suffolk where Lady Eve Balfour started her comparison of the results of mixed, stockless and organic methods in 1939 and where the experiment continued until 1969. In my opinion this was a most valuable study, and I regret that the results have not received greater scientific evaluation. I am convinced there is something to evaluate. The results indicated that the nutritional value of the organic produce was different, and, when fed to cattle, superior to that from the other sections. It, and the work on other organic farms, has shown that this system can be a viable alternative to the orthodox system, and if the superiority of the produce can be established it, or at least some of its ideas, are likely to be more widely adopted.

In this book I am only concerned with the possible effects of organic farming on wildlife. Even a cursory survey of organic versus traditional farms will show that birds and other organisms are more plentiful on the organic farms. However, I do not think that this is entirely due to the farming system. Almost all organic farmers are sympathetic to wildlife, and therefore keen to leave features on their farm which may encourage its survival. They remove fewer hedges (though those with large farms have removed sufficient to ensure efficient working), they keep their livestock out of doors, they retain at least some unimproved grass, believing that this makes a special contribution to animal nutrition. They mostly do, off their own bat, just those things we are trying to persuade all farmers to do to obtain a compromise between efficiency and sterility in the countryside.

There is more doubt as to whether, on the areas on which crops, particularly arable crops, are grown, the organic method makes a great deal of difference to wildlife. More reliance is placed on nitrogen fixation by leguminous plants. Thus a cereal may be undersown with trefoil which is left growing over the winter after harvest, to be ploughed in in spring before the next cereal crop is planted. This will give food for wildlife all the winter through, grain from the stubble and green matter from the trefoil. The reduced use of chemical fertilisers will reduce the danger of damaging eutrophication in near-by water, and if no toxic pesticides are applied there is no danger of poisoning wildlife. Otherwise large-scale organic farming, which will often have to be intensive to make it pay with land so expensive a commodity, will have much the same effect on the landscape as farming using chemicals. So the adoption of the organic method by British agriculture would not be a panacea for good wildlife conservation, though it might make it just a little bit easier.

CHAPTER 3

GRASS AND GRAZING

TODAY grass and rough grazing cover well over half the land surface of Great Britain. Official statistics suggest that we have about two and a half million ha of temporary leys, productive swards sown by the farmer and, after a few years, ploughed again to grow arable crops. We have twice as much permanent grass, and seven million ha of rough grazing. In addition substantial areas of golf courses and sports fields are covered by turf, and there are at least one hundred thousand ha of mown grass in private gardens. Botanical studies show that some of the so-called permanent grass is in effect re-sown ley, and that all these data may require modification, but they indicate the order of magnitude of the different types of land use. There is a very large area of grass of one type or another. Most of this grassland was once forest, and left to itself it would return quite quickly, in 20 or 30 years in most cases, to the same state. Grassland has been created by man in order to feed his grazing livestock. Yet it may be prized by conservationists, though its protection and management present many difficult problems.

Grassland in Britain takes many different forms. Different species of grass and other plants are found in different areas depending on the nature of the soil, the altitude and the rainfall. The most 'interesting' grasslands are generally those which have developed slowly, perhaps over hundreds of years, and which consist of indigenous plant species. Although the ground was originally cleared from forest by man, and although shrub and tree growth has been prevented by grazing by farm stock, such areas are accepted as at least 'semi-natural'. Most of the plants which persist have not been introduced by man. The vegetation may bear some resemblance to the small patches which existed in forest clearings before man became a farmer. When, perhaps three hundred years ago, chalk grassland covered much of the nearly one-and-a-half million ha of soil based on the chalk deposits of the Cretaceous period, plants characteristic of this soil must have been many times more common than they were before the forests had been cleared. With the increasing destruction (or 'improvement') of this type of grassland, these plants are, once more, becoming rare. Similarly the flower-rich meadows found in the lowlands, and the upland pastures with their characteristic flora, produced a temporary abundance of many plants which, both in the original natural woodland and in the fields of the modern farmer, existed only in much smaller numbers. Our attempts at grassland conservation must therefore be recognised as being aimed at encouraging species and associations which we value for scientific or aesthetic reasons, not because they are natural. The problems involved in conserving different types of grass are complex, and need to be considered in some detail.

A temporary grass ley is essentially an arable crop. It may, when first planted, contain no truly indigenous plants. Various seed mixtures have been used in the past, and are being used today. The tendency has been, in recent years, towards simplicity, using seeds of only a very small number of species, and these are either entirely foreign (e.g. Italian ryegrass) or improved cultivars of British species. The goal has been to obtain rapid establishment of a sward which may then be grazed; some leys in which ryegrass is predominant are ready, in favourable seasons, in six weeks. Emphasis has also been on high productivity, and the ability to react efficiently to heavy dressings of chemical fertilisers, particularly of nitrogen.

Clearly such a ley is of little interest to the botanist. The better it is managed, the less any interest develops. However, it is generally impossible to prevent some native weeds from appearing. Thus within two years of planting a mixture of perennial ryegrass and white clover a substantial number of plants of other grass species, particularly Yorkshire fog (*Holcus lanatus*) and bents (*Agrostis* spp.) as well as a number of broad-leaved species, are likely to appear. But the result will be quite unlike that which appears over much longer periods in old grassland. Good farm management is aimed at reducing the number of weeds in the ley.

Some farmers have thought it worth while to include more species in their leys. They have planted deep-rooting herbs like chicory which bring minerals to the surface from the sub-soil. There is some evidence that herb-rich leys, generally favoured by organic growers, do contribute to the nutrition of grazing stock, though most farmers prefer to supplement a more conventional diet from a high yielding ley, even if this lacks in quality what it provides in quantity.

The longer a ley persists, the greater the number of species ('weeds') and the greater the resemblance to old grassland found on similar soil. However, it seems probable that it would take tens, or even hundreds, of years before a chalk area became the host of all the plants found on the Wiltshire downs where no cultivation or planting has occurred for millenia. As most leys are replaced by corn or other crops after two, three or four years, this invasion is kept to a minimum. Leys have their value to the farmer, they act as a break when weeds and diseases of arable crops may be brought under control, they allow a build up of organic matter in soil which has been depleted, and they provide fodder or grazing for livestock. Pigeons, deer and rabbits may share the grazing with the farm animals, and the earthworms and other soil animals may flourish better than in soil ploughed annually, but the contribution to wildlife is obviously minimal. There is no way in which ley management could be modified to make these areas much more valuable to conservation. Too great a use of fertilisers may cause eutrophication by run-off into streams in the vicinity, and attempts to control grass pests could harm non-injurious insects, but reduction in the use of chemicals will have little positive effect and conservationists must accept the fact that short-term leys cannot be made to contribute substantially to enriching the countryside, though they may help to keep the soil healthy and productive.

I have already (chapter 1) pointed out that arable farming first developed on the chalk downlands, where trees could be easily felled and the thin soil cultivated. Only a few successive crops were taken from any one area before the fertility was reduced, and the area was then allowed to revert to grass. Cattle and sheep grazed the grass, and gradually the flower-rich turf developed. This provided good grazing by the standards of the day. It also provided much interest to botanists – the literature on chalkland ecology far exceeds that on all other grasslands combined.

The grass downs were exploited as sheep walks for many hundreds of years. It was common practice to graze the animals by day, and then to enclose them overnight with hurdles on the arable fields, which they improved with their dung and urine, and by trampling the soil with their hooves. However, this system began to be less important three hundred years ago, when new crops enabled more livestock to be kept so that more farmyard manure was available. The areas of chalk grassland then decreased. This process has continued, and today only about three per cent of the chalk has a cover of grass, and some of this has been improved out of all recognition by the use of herbicides and fertilisers.

When chalk grassland is destroyed, it has probably gone for ever, for it has neither the time nor the opportunity to regenerate under present farming pressures. In the days when large acreages were under grass, a small patch which had been ploughed and cultivated would afterwards be re-colonised by chalkland species from the surrounding grass and flowers, thereby replacing those that had been lost. Today there is quite a number, though a rapidly decreasing number, of isolated relict fields of chalk grassland. If one of these is destroyed the only seeds likely to colonise the ground are those from the surrounding land, so the chalkland species disappear.

Eventually, with suitable management, chalk grassland may be reconstituted, but this may take a long time. Experiments involving the planting of typical species have not yet been entirely successful. The whole process appears to be very slow. One of the characteristic flowers of the chalk is the Pasque flower (*Anemone pulsatilla*). Once occurring locally in considerable numbers, it is becoming increasingly rare as its habitat is altered, and few new seedlings seem to establish themselves. Thus there are many railway embankments in chalk areas, and these have now existed for well over a hundred years. The ground has been colonised by grasses, some flowering plants, shrubs and even trees. Some useful wildlife areas have resulted. However, there is no record of the Pasque flower establishing itself on a railway embankment; the competition from other species seems to have been too great.

Chalk grassland has been lost by being intensively farmed, but it can also disappear when neglected. The herb-rich turf was established under quite hard grazing, particularly from sheep. Plants like the Pasque flower could survive even very heavy grazing. The flower heads were often bitten off when still in bud, but the rosettes of leaves at ground level survived, and the species reproduced vegetatively. When grazing in spring was restricted, masses of purple flowers of *A. pulsatilla* appeared, and also the blossoms of many other species which were not apparent when the grass was grazed. This gave the impression that greatly reduced grazing was beneficial if good flower cover was desired.

However, we have learned that if chalk grassland is to be conserved it must be subject to some grazing. In the 1930s British agriculture was badly depressed, and the number of sheep grazing on the downland decreased greatly. At this same time rabbits became very numerous in these areas, and they replaced the sheep as grazers. The turf remained similar in appearance and supported the same plant species as when it was regularly grazed by sheep. Then in 1953 myxomatosis arrived in

Britain and almost wiped out the rabbits. Few sheep remained on the chalk. For a couple of years the chalk grassland was covered with blooms. Then coarse grasses started to accumulate, shrubs such as hawthorn appeared, and the number of flowers decreased. It was apparent that the area would soon revert to scrub and then to forest, and that the grass and flowers would disappear. We know therefore that if we wish to conserve chalk grassland, we must continue sheep grazing in the traditional manner, or, as a second best, we can mow the grass at the appropriate time. The conditions we now wish to conserve are those which were produced accidentally by one, now obsolescent, type of farming practice.

What chalk grassland remains has survived largely by accident. Much is in Wiltshire. Some is on steep banks which are difficult to cultivate but where sufficient grazing still takes place. Nearly half is on land held for training by the Ministry of Defence. There are many complaints that the army is keeping the public from beautiful parts of the countryside, but in this and other cases military occupation has prevented further exploitation and been a positive gain for conservation. As the army does not have to make its living off the land, it can still permit grazing levels which have optimum effects, and there is no need for pasture improvement to allow much higher stocking densities on the more productive but species-impoverished sward. But in the long run more nature reserves with grazing controlled for the benefit of the plants and the animals living on and with them is the only sure way of ensuring the survival of this habitat. For instance, insects may be dependent on the appearance of flowers or on the ripening of seed, and a mosaic of hard grazed, lightly grazed and ungrazed areas (something difficult to obtain on a commercial farm) may be necessary if a wide range of species is to survive.

A whole series of different types of grassland is familiar to botanists. There are the other calcareous grasslands on the different types of limestone. Large parts of Britain supported neutral grasslands, including washlands in East Anglia and flood meadows in the Thames valley. The most widespread areas of semi-natural grassland occur on acid soils, and include sandy heaths in Dorset and upland pastures in Wales and Scotland. There are salt-marshes, sand dunes and other special categories with their characteristic flora and fauna. They all have one thing in common: their area is decreasing year by year, and where they remain, modern farming practises are usually making them much more productive, and able to support far greater numbers of livestock.

The farming press today is full of success stories of pasture improvement. I have a selection of extracts before me. 'Drainage turns rough pasture into corn land' describes how poor grazing land on the edge of Dartmoor was improved, so that it now yields more than five tonnes of barley to the hectare. 'The road to reclamation' describes how poor grass in the Welsh hills was rotavated and left to rot for 18 months. then a pioneer crop of rape was planted, which in turn made way for permanent good quality grass and clover, providing food for thousands of sheep and hundreds of cattle. 'Stocking rate soars on improved hill land' describes what happened on a large farm in Angus in Scotland. Much of the land was originally covered with what the writer calls, somewhat contemptuously, 'natural vegetation'. This was mainly heather and nardus grass, with patches of broom, gorse and bracken. The land was ploughed, cultivated, fertilised and reseeded and supports, in summer, 675 fat cattle where previously only 60 could obtain a poor living. Many of these stories are illustrated by excellent colour close-ups of the weedless grass and luxuriant clover now replacing the flower-rich turf which was so much less nutritious.

Those familiar with different parts of Britain will have seen this process in action. As a boy I knew Upper Teesdale well. Above Middleton in Teesdale there was little arable, but wild flowers, of which the spring gentian (*Gentiana verna*) is the most famous, abounded in suitable localities. On the damper grazing meadows the globe flower (*Trollius europaea*) was common in early summer. The mealy primrose (*Primula farinosa*) grew freely on drier ground. In 1979 I went on a farm walk on fields I knew fifty years ago. The management of the grass was excellent. Cattle and sheep grazed at a density which would have been unthinkable in the past, and hay and silage of the highest quality was made from several fields. Careful use of lime and basic slag, with the return of all manure to the soil, ensured a steady rise in fertility and the continuation of top quality pasture. The soil fauna clearly flourished under the grass. But the beautiful flowers I remembered so well had all disappeared.

Incidentally, it may be possible to get an over-optimistic view by looking at some official statistics. The Annual Returns for 1971 for Huntingdonshire showed the very respectable area of 13,545 ha of permanent grass and rough grazing – this is roughly 14 per cent of the total area of the county. A careful survey showed that only 356 ha (3 per cent) were of any botanical interest, the rest had been improved by the use of herbicides and fertilisers. This story could be repeated for many other parts of Britain.

One traditional form of management was the water meadow. Low lying flat pastures along river banks were flooded using a complex system of hatches or sluices and artificial channels. The meadows were usually drowned for a period of two or three days in winter, between December and late March. The purpose of the flooding was to warm the soil and prevent it freezing. Often a layer of nutrient-rich silt was deposited on the surface. The meadow produced an early flush of grass, which was of particular value to ewes and lambs at a time when food was scarce and plant growth in other fields was slow. The meadows were also flooded in times of summer drought. Although some flowering plants such as the primrose (*Primula vulgaris*) and the cowslip (*Primula veris*) are generally killed by flooding, others, such as the fritillary (*Fritillaria meleagris*), flourish. Water meadows generally showed considerable botanical diversity.

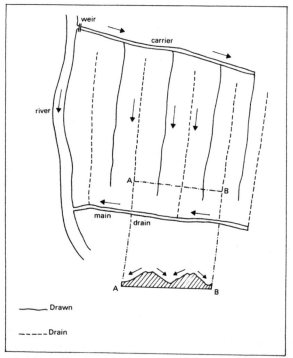

FIG. 3. Plan of the layout of a water-meadow, showing the channels used for drainage and irrigation. (After Duffey et al 1974)

Although these fields gave a valuable early bite, and produced grazing and hay of above·average quality and quantity, the operation of the flooding regime has been abandoned as the sluices were difficult to maintain and the whole process was labour intensive. Most of the water meadows have been reseeded and heavily fertilised. The total yield of herbage has thus been greatly increased, and the additional hay and silage, which can be stored through the winter, more than compensates for the loss of the early growth on the water-warmed land. One improved area was found to contain little more than perennial rye grass and timothy, instead of the beautiful and varied sward which previously existed.

The pace of grassland improvement has increased in recent years. Until 1945 most grass received little fertiliser, except lime, basic slag, and limited amounts of farm manure. The greatest addition was the dung and urine deposited by animals while grazing. Manure increased herbage productivity, but did little to change the composition of the sward. When first used, chemical fertilisers, particularly nitrogen, gave spectacular increases in grass. However, as the total needs of the soil and of the plants growing on it were not always understood, long-term results were often disappointing. Yields fell off, soils became acid from too much ammonium sulphate, and productive grasses were replaced by inferior species. Today, as the result of research at such centres as the Welsh Plant Breeding Station at Aberystwyth, the Grassland Research Institute at Hurley in Berkshire, Rothamsted, Seale Hay Agricultural College in Devon, and many practical far-mers, results can be foretold and damage avoided. Modern pasture produces better and better grass year after year, and its soil becomes increasingly stable with enhanced fertility. Unfortunately native plants and animals generally suffer.

It is widely believed that many species of butterfly are rarer in Britain today than they were fifty years ago. Frequently toxic in-secticides are blamed, but this is probably of comparatively little importance (p.124). Individual insects may be killed when crops are sprayed, but there is no evidence of a butterfly population being harmed by these chemicals. The species which seem to have been adversely affected are those whose caterpillars feed on native grasses and on the herbs of unimproved pasture. Both rare and common species are involved. Thus one widely used handbook published in 1905 describes the meadow brown (*Epinephele ianira*) as 'our com-monest butterfly', and it was certainly very abundant when I was a child. Though still numerous in some areas, it is greatly reduced in others, and this reduction is undoubtedly caused by the loss of old grassland. Other species which have been affected are the adonis blue (*Agriades thetis*), which depends on the horseshoe vetch (*Hippocrepis comosa*), the small blue (*Cupido minimus*), whose caterpillars feed on the kidney vetch (*Anthyllis vulneraria*), and the silver spotted skipper, whose food plant is the grass *Festuca ovina*. Other butterflies from wetlands and woodland rides have been similarly affected. However, the only actual breeding species which has been exterminated in recent years is the large blue (*Maculinea arion*). Here again habitat loss is mainly responsible. This is rough grazing land where the food plant, wild thyme, (*Thymus drucei*) grows and where anthills abound. The later instars of this caterpillar feed on ant larvae within anthills.

Changes in the timing of farming operations have also had their effect. Grass can provide shelter and nesting sites for birds, as well as food for many species. Grazed fields provide few nest sites if most of the grass is removed early in the season. Fields shut up for hay were in the past not cut until July; these provided safe havens for a host of birds, insects and other animals. Today improved grassland produces a hay crop which is ready to cut many weeks earlier than were grass fields in the past. Furthermore an increasing area of grass is cut several times in the season, starting in early May, in order to produce silage for winter feed. This early cutting is harmful to wildlife, particularly ground-nesting birds. It has contributed to the reduction in numbers of the partridge, and is probably responsible for the elimination of the corn-crake (*Crex crex*) as a breeding species from most parts of Britain. It was once more common. I remember in my own schooldays in Teesdale how we disliked this bird, for its noise from hayfields outside the dormitory window seemed to go on all night. Today the same fields harbour no corncrakes, whose breeding is now confined to a few remote sites in the Scottish highlands and islands.

I have so far been mainly concerned with lowland grassland, almost all of which is farmed, and most of which has a rapidly decreasing value for the conservation of our native flora and fauna. We also have the huge area of seven million ha of 'rough grazing'. In fact there is often no rigid demarcation between the poorer (from the point of view of food production) permanent grass and the rough grazing, and also we get a gradation between such areas and those thought of as heath or moor. It is possible to generalise and say that the better the soil, and the less extreme the climate, the greater is the pressure on such areas from modern agriculture. In some of the least productive areas pressure is actually less today than in the past; unfortunately this is not always a gain for wildlife.

As has already been stated, most of Britain, except the mountain tops, was originally covered with forest, and if left alone forest would return. There is evidence that 10,000 years ago birch woods extended over 900 m (3,000 ft) but, possibly because of climatic change, the tree line today is between 450 and 650 m. Above that some hardy trees may survive and grow slowly, but the montane flora consists mainly of shrubs, increasingly prostrate as the highest levels are reached, grasses, mosses and lichens, with considerable areas of bare soil and rock which may be gradually colonised by the slow growing plants. At these high levels wild animals are comparatively scarce. Some creatures are adapted to living there, for instance the rare dotterel (*Eudromias morinellus*) seems to shun nesting sites below 900 m in Britain (though recent colonisation of polders below sea level in the Netherlands by this bird is something of a nightmare to tidy-minded ecologists).

These highest areas are probably the nearest we have in Britain to ground and vegetation in its natural state. Man has had some effect. In the Cairngorms ski lifts take more people than before to the highest levels. Increased erosion and damage to plant life has occurred. There has been some effect from grazing sheep, but seldom on the mountain tops. Near to and below the tree line the effect has been more serious.

There is growing pressure to improve the better areas in the uplands. As has already been mentioned, corn is now grown successfully on what was until recently moor in Devon, but most effort is devoted to improving pasture for cattle and sheep. One change which has been welcomed by many conservationists is the better control of bracken *Pteridium aquilinum*. This invasive plant is found on the better soils where more valuable species are preferred. It has little economic value except perhaps as bedding; it harbours flies and can make walking almost impossible.

Until recently bracken could only be eradicated by frequent cutting and special cultivation techniques. These could not be applied on rough ground where the bracken continued to flourish, and from which cleared patches were quickly reinvaded. Now the herbicide asulam has been found to be very effective. Properly used, the chemical does little damage to other forms of wildlife (p.112). The only concern is that it is so effective that aerial spraying could make unacceptably large changes in the uplands. Although too much bracken is generally deplored, it would be a pity if it were eliminated entirely.

For many years the more productive areas in the uplands were seriously overgrazed by sheep. This affected the vegetation, reducing its productivity and its botanical interest. Sheep are selective feeders, and may kill the more palatable grasses by eating them right down to the roots. Today the tendency is to devote most efforts to improving the best areas, and to reduce the use of the poorer. Thus one report says: 'The viability of farming in the uplands is precarious, and evidence of reduction in management can be seen in abandoned buildings, collapsed walls and reverted pastures in many areas." Some more remote and mountainous moors are becoming covered with scrub and the early stages of forest regeneration. This is a development which some will welcome if it does not affect too great a proportion of the moorland, but it could reduce diversity and eliminate many desirable plants.

In many areas traditional methods continue and are being improved. Much of our moorland is covered with heather, and for this to provide grazing for sheep and grouse, it must be burned at intervals to get rid of shrubby material and to stimulate palatable and nutritious young growth. The official guide on the subject of what, in Scotland, is called 'muirburn' states that it is 'a means of managing the vegetation of upland grazings and grouse moors to maintain it in a productive condition'. The law prohibits burning except between 1 October and 15 April. The guide suggests that small patches, $\frac{1}{2}$ to 1 ha, should be burned in a rotation of perhaps 10 years, to produce a mosaic of different heights of heather. Advice is given on choice of day, weather and labour needed to ensure the best results.

Scottish moors are burned in many cases primarily for the benefit of the grouse, but most other wildlife may benefit. Some plants may be destroyed, and animals killed, by the actual process of burning, but little damage and much good is done by adhering to the principles agreed by the Department of Agriculture and Fisheries for Scotland and the Nature Conservancy Council. Here we have a case of a traditional method which has been in use for many years continuing to be used, but with some improvements based on recent research. So long as grouse shooting on the Scottish moors remains big business and keepers can be persuaded not to kill every predatory bird they see, much of the other wildlife of these areas has a reasonably secure future.

The greatest 'danger' to much of our uplands is not modern farming, but afforestation. There is considerable opposition to the plans of the Forestry Commission and the private forestry industry to replant several million hectares of the uplands to try to make Britain less dependent on timber imports in the next century. Were replanting to be restricted to native trees, most conservationists would approve, but they disapprove of the introduction of more productive exotic conifers such as *Pinus contorta* and spruce, *Picea sitchensis*. Farmers also fear that they would lose much grazing land, though it seems likely that the increased shelter from the trees, with improvement of the remaining land which could be integrated with the tree planting might actually increase the possible stocking capacity and make better farming with improved supervision of livestock possible. The only disadvantage would be to the flora and fauna. Birds and other animals adapted to open moors would decrease. It is likely that the golden eagle (*Aquila chrysaetos*) would be adversely affected, in this case because fewer carcasses of lambs and sheep would remain to rot on the moors and an important food source of this carrion-eater would be removed.

One interesting grazing area which is being lost is the salt marsh, as it is being drained and reclaimed to grow arable crops. This is considered with other disappearing wetlands in chapter 7.

One final point may be made about grassland management. As a rule frequent hard mowing discourages many types of wildlife, particularly if the area is to be used by breeding birds or timid mammals. However, even the shortest turf may provide feeding grounds for many species. Suburban householders can observe the large number of birds which feed on their lawns. Starlings are particularly numerous, probing the turf to remove leatherjackets, the larvae of the crane fly or daddy longlegs, *Tipula paludosa*. A lawn, a playing field or a golf course may provide a substantial supplement to the diet of the birds, and allow more of them to survive than could be supported from more natural and less well groomed areas of turf.

CHAPTER 4

LIVESTOCK

THERE have been great changes in the way farm animals, cuttle, sheep, horses, pigs and poultry are kept on British farms, and these changes have had their effects on wildlife. As with arable crops, the yields of animal products from British farms have increased greatly in the past forty years. As compared with 1938, in 1977 milk and eggs had doubled, as had beef and pig meat. Mutton and lamb had only increased by some 10 per cent, but poultry meat had increased from 90,000 to 679,000 tonnes, a factor of seven and a half. As a result roast chicken, a luxury food before the 1939 war, is the cheapest meat in Britain today. We eat on an average over half a pound (300 g) per head every week.

In 1977 there were 13,854,000 cattle, 28,104,000 sheep and lambs, 7,736,000 pigs and 133,886,000 head of poultry. The sheep were only slightly above the pre-war figure of 25,786,000, but in all other cases the population had roughly doubled. These gross figures may be misleading, particularly in the case of the poultry. The number quoted for 1977 is for the birds alive on a particular date; the weight of poultry meat produced was so much greater as most of the birds are now ready for the table in 8 to 10 weeks, whereas in 1938 many were not eaten until they were six months old.

It is the objective of the livestock farmer to produce the greatest possible amount of meat or animal products from every hectare of his farm. He has no wish to share the grass with wild herbivores, nor is he prepared to sacrifice his livestock to wild carnivores. Thus he tries to eradicate rabbits, and though he may even encourage deer in small numbers, particularly if he can shoot a few for sport, venison or the continental meat market, he will not tolerate deer if they obviously compete for grazing with his farm animals. Our ancestors got rid of the large carnivores like the wolf, and the smaller species which survive often do so against the odds. Foxes are persecuted in areas where hunting does not take priority, and eagles are still shot by those who believe that they carry off lambs to feed their young.

The earliest graziers were nomads who drove their beasts over the countryside in search of food. They only kept as many animals as they could feed off the natural vegetation, for which they competed with wild herbivores. In good years stocks increased, but when droughts occurred many died of starvation. When farmers settled down and grew cereal and other crops, livestock was accommodated on poorer land and also fed on crop wastes. Later special crops like turnips and clover were grown to feed farm animals, and their numbers then increased far above those which could be sustained on natural vegetation or even on fresh and conserved grass. The use of grain and protein foods has enabled even greater numbers to be maintained. Some livestock (e.g. pigs and poultry) may feed only on material bought in from other farms or even from other parts of the world. These different ways of feeding and managing livestock all affect wildlife and the countryside generally.

In the wild, animals of different species exist in the countryside in numbers which can be supported by the food available. We have a very diverse landscape, with many different plants and animals, and the numbers of the different species remain very approximately the same year after year. There are indeed fluctuations, but increases to pest proportions, with obvious damage to the habitat, are uncommon. The diversity which exists under these natural conditions produces this relative stability. The stability is far from absolute, but the situation is very different from that of a modern farm, with rapidly changing cropping patterns and sudden switches from livestock to arable, or from dairy farming to fattening bullocks.

Wild grazing animals, be they antelopes in Africa or caterpillars in Britain, seem to be present in most cases in numbers far smaller than the food available at some times of year might support. We may get the false impression from television that wild herbivores cover the plains of East Africa in enormous numbers, and that they are crowded together quite uncomfortably. This is only true when we are shown animals protected on nature reserves, for the general stocking densities over wide areas are very low. This can be demonstrated by looking at the herbage, which is seldom cropped down to the ground so that at the end of the growing season many patches of tall grass with flower heads remain. This then forms the food for other creatures, and eventually is broken down by fungi and bacteria and its nutrients are recycled to feed more growing plants. On a farm, however, the grass is either grazed as hard as possible, or, if there is a surplus, this is cut and conserved as hay or silage.

Until 1939 most livestock on British farms lived out of doors, grazing on permanent grass, for most of the year. The grass was not used with maximum efficiency, as the animals had access to the whole of a field, even if it was not always a very large field, and much was damaged before it could be consumed. The grass therefore tended to be grazed rather unevenly, patches soiled with dung being left, and trodden areas were not completely eaten. In wet weather muddy worn patches were produced, which added to the diversity of the area. As little fertiliser and no herbicides were used, the floristic diversity discussed in the last chapter was maintained. Cowpats provided food for dung beetles and other insects, and the tufts left ungrazed provided food for many forms of wildlife. A large population of flies was produced, providing food for bats and other insectivores.

One change which has had considerable effects is the reduction in the number of mixed farms, particularly in Eastern England. Fifty years ago almost all East Anglian farms supported considerable numbers of livestock, as well as growing arable crops. Most farms were, for that time, highly productive, and this was demonstrated by the fact that they employed more labour than did farms in other parts of Britain. Today a great many of these farms are one hundred per cent arable, with no livestock at all. Farm employment is now lower than in other areas. It is possible to travel many miles without seeing a single grazing animal. There are still pigs and poultry, but these are kept in intensive indoor units.

Where livestock is still kept everything is done to ensure efficiency. Productive grass leys are often grazed one narrow strip at a time, the rest being protected by an electric fence until required. No uneaten patches are left for other creatures. 'Zero grazing', which means that animals are kept indoors and their food is cut and brought to them, is also practised. Grass is cut for silage long before it is mature to produce high quality, high protein feed, so many fields look like mown lawns for much of the season. This may encourage birds like starlings which have easy access to insect grubs in the soil, but the richness of the wildlife is greatly reduced. More traditional (? old-fashioned) animal husbandry has not entirely disappeared. There has been less change in the West of England and in Wales, but even here farms are getting larger, more intensive methods are being introduced, and a greater part of the output of the land is going to the livestock and less to support wildlife.

Perhaps the most dramatic change has been the disappearance of the working horse from British farms. These animals at their peak, in 1910, numbered 1,137,000, when they were the only source of power, except for traction engines ploughing heavy clay and threshing corn. There were still 650,000 working horses in 1939, but only 161,000 in 1945. Bullocks as draught animals had also been common until well into the nineteenth century, but virtually disappeared except for one team still at work for an enthusiastic owner. Today there are still a few keen supporters of the different breeds of horses – Shires, Suffolk Punches, Clydesdales, and these may occasionally still be seen at work – but on 99 per cent of farms the internal combustion engine is the sole means of ploughing and harvesting. The disappearance of the working horse has meant that the two million acres of grass and arable needed to feed them in 1939 is available for other uses. The overall effect on wildlife has probably not been very great, except that horses, being selective grazers, leave diverse herbage which may support different invertebrates (see p.55).

Rising fuel prices and future oil shortages have made farmers look into the possibility of bringing back working horses. Any major re-introduction is unlikely, as horses use too much energy. Unlike tractors, which only use fuel when actually at work, they have to be fed every day. The food needed by horses is, in terms of energy, many times greater than that of oil used by tractors to do the same work. Further-more, to harvest present-day crops with slow-moving horses would be impossible – it is difficult enough in bad weather with rapidly operat-ing combine harvesters. There may be some use of horses on small farms with spare rough grazing to supply free food, and where stock is kept on the hills 'cowboys' on horseback may be more efficient than landrovers or trail motorcycles, but these are likely to make only a minor effect on national statistics.

However, the horse in Britain is by no means facing extinction, as we have over 500,000 animals kept for recreational riding, and 11,600 racehorses in training. There are many undergrazed paddocks with considerable wildlife interest, and the many racecourses provide large areas of short turf on which wildlife can feed.

Intensive livestock enterprises – 'factory farms' if you do not like the system – are increasing. Many people find these unattractive, and object to the ways pigs, veal calves and poultry are now kept. Farmers retort that these objections are based on misunderstandings of animal psychology, and that as the animals grow well and appear healthy the situation is better for the livestock than shivering out of doors in winter.

This is not the place for any discussion on the ethics and economics of these systems; I am solely concerned, in this book, with the effects on wildlife.

As I have already indicated, animals grazing in the field may favour some forms of wildlife, though, unfortunately, the less efficiently they are maintained the better do wild creatures survive. The removal of animals to indoor units is therefore in itself harmful to wildlife. Quite apart from this, intensive enterprises may have their own effects, almost all harmful, on wildlife. The Royal Commission on Environmental Pollution, in its report on Agriculture and Pollution, published in September 1979, stresses the danger arising from the improper disposal of excrement from livestock kept intensively. While farm yard manure has, traditionally, been the major means of maintaining soil fertility, excrement from animals in factory farms has often proved difficult to dispose of, and has been a serious source of water pollution. Pig and poultry farmers, feeding animals on bought-in food, often have no land on which to spread their manure. Slurry from cattle or pigs smells objectionably and is too liquid for many purposes. Some manure from pigs contains copper (used as an additive to animals living in isolated houses) which renders it phytotoxic. Some farmers have lagoons where the material is stored; these may overflow in wet weather and contaminate rivers. Some wastes are treated, expensively, in sewage treatment works. At one time much was discharged, untreated, into rivers, but this has now virtually stopped.

Nevertheless animal wastes still cause pollution and damage the wildlife, both plant and animal (including fish) in rivers, lakes and streams. Damage can be done by the excessive amounts of organic matter, or by nutrient salts, remaining after treatment giving rise to eutrophication. This term is often used as if it were a synonym for water pollution, which it is not. It really means good nutrition. Unfortunately we do not normally like the results of such good nutrition, and prefer the type of life which exists in waters containing low levels of nutrients. Thus trout live best in badly nourished (oligotrophic) rivers. Most natural rivers in the hills are oligotrophic, and they pick up nutrients as they flow through low-lying land on their way to the sea. The change is gradual, and life in the water is adapted to this. Man's efforts may put too much nutrient into water too quickly. This encourages algae, which blanket the surface. When the algae blanket decays, it may deoxygenate the water, killing fish and other animals.

Our Water Authorities are successfully reducing the pollution and

eutrophication caused by farm wastes. Rising costs of chemical fertilisers are making the use of animal manures more economic, even
when the material has to be transported from a livestock unit to an
arable farm. In the long run I think that the harmful effects of this type
of pollution will be almost completely eliminated, but at present they
are still appreciable.

While many people are distressed by the spectacle of hens in battery
cages, the cages do at least keep predators out. It is rare, today, for
chickens to be eaten by foxes. Actually domestic poultry were never an
important part of the diet of the fox, which lived largely on small
mammals, worms and other invertebrates, and on carrion, but great
slaughter could occur when a fox broke in to an old-fashioned hen
house. Occasionally similar slaughter occurs in deep litter houses or
buildings containing broilers, but as a rule these are too well con-
structed to make such entry practicable. Poultry today is not an
important food for wild carnivores.

Man has been rather unimaginative in the way he has only domesticated a small proportion of the world's mammals, but attempts are
now being made with other species. Today work is in progress in
Britain to try to domesticate our largest truly wild animal, the red deer
Cervus elaphus. Deer stalking as a sport has a long history, and care is
taken on well-run estates to cull surplus individuals to keep the numbers within the optimum carrying capacity of the land, though in very
severe weather hay may be fed to offset the danger of starvation (and,
in some cases, to discourage the deer from straying onto arable crops
and damaging them). This is accepted as good wildlife conservation.
Now attempts are really being made to domesticate the red deer which
it is thought may be economic as it can utilise different, and poorer,
grazing than can sheep which now tend to monopolise many upland
pastures. If this work succeeds it will transfer the deer from the ranks of
wildlife to those of domestic livestock. As it may be necessary to remove
the antlers of stags kept confined in paddocks to prevent fighting (and
to sell the young antlers in velvet which some believe to have aphro-
disiac properties) or to castrate the males when they are young, these
will not have the romantic appeal of free-living stags, though they will
be easier to see than on the moors, and deer farms may then become
tourist attractions.

The red grouse (*Lagopus scoticus*) has the distinction of being the only
truly endemic British bird, naturally occurring only in the British Isles.
Moreover, grouse shooting is an important industry, and is responsible

for the way in which large upland areas are managed. The grouse grazes mainly on heather, which is managed by 'muirburn' in the way mentioned in the last chapter mainly for the grouse's benefit, though sheep also profit from the young growth which proper burning stimulates. It is most unlikely that grouse will be reared intensively and, so long as they provide most income from the area in which they flourish, the traditional methods of land management, which maintain the native flora and fauna in a generally satisfactory manner, will persist.

I must now return to our more traditional farm animals. These have for many hundreds of years lived under unnatural conditions, but until recently these conditions were those where at least some of our wildlife flourished. Modern practices, if they are to be economic, are increasingly harmful to wildlife. Now serious attempts are being made to use livestock to aid wildlife conservation. This means using regimes which do not have the maximum productivity, and which therefore may not be financially viable as farming practices, yet they may contribute substantially to food production and as conservation methods, they are very economical.

One interesting experiment was made at Woodwalton Fen in Huntingdonshire, a National Nature Reserve for which I had, at the time, the management responsibility. Before the fens were drained, although they were drowned during much of the winter, they became drier during the summer and large areas were extensively grazed by cattle. When Woodwalton became a nature reserve, such grazing ceased. Until 1954 rabbits were common and kept the grass short, but then myxomatosis wiped the rabbits out in this area. The coarse grasses *Calamagrostis epigejos* and *Agropyron repens* spread rapidly, and appeared to be smothering the rest of the vegetation, which included various interesting species, among them the uncommon Deptford pink (*Dianthus armeria*), the marsh orchid (*Dactylorchis incarnata*) and the heath violet (*Viola canina*). We decided to try to reintroduce cattle grazing. At first some Friesian bullocks were borrowed from a neighbouring farmer, but then we decided to buy our own stock so that we could have full control over them. Otherwise there could have been conflict, as the farmer might have wished to graze the land just when we wished, for conservation reasons, to reduce this pressure.

We decided that we wished to use cattle which could overwinter under the bleak fenland conditions, where there is nothing to provide shelter from the cold east winds from the Russian steppes. I always regretted that we decided against Highland cattle, which would have been picturesque as well as hardy. We decided on Galloways, which were almost as attractive, and introduced ten steers, approximately 12 months old, weighing 195 kg each, in 1966. These were kept on the fen for two years. During the winters they barely maintained their weights, and had to have their diet supplemented with small amounts of concentrates and a bale of hay among the ten steers each day. In summer they gained well on the diet of coarse grasses, *Phragmites* and *Juncus*, reaching nearly 500 kg before being sent to market. They were then older than animals of similar weight kept more intensively, but they aroused much interest particularly among the older butchers and sold at an excellent price. The flavour of the beef was outstanding, quite unlike the tasteless product from intensive 'barley beef'. Though the return would have been insufficient for a commercial farmer paying wages and rent, it was, from the standpoint of food production, not a bad way to use a nature reserve.

This experiment succeeded in its main objective. The coarse grass was largely removed, and the Deptford pink increased. In one area where only five flower-bearing individuals were found in 1964, these increased to over 900 in 1967, and a similar improvement occurred with many other species.

More sophisticated grazing regimes have been adopted on other reserves, for instance using sheep on chalk grassland. Here it has been found that Border-Leicester × Cheviot (Scottish Half-bred) are a suitable breed, if they are to overwinter without receiving supplementary feed, except perhaps in the most severe weather. The Nature Conservancy carried out a series of trials using different densities of sheep on different areas during the growing season. They compared the effects of sheep and mowing with mechanical means. It was found possible to produce a mosaic of patches with grasses and herbs at different heights and states of development in each. This sort of treatment would obviously be uneconomic on an ordinary farm, but on a nature reserve sheep are a cheap and efficient method of mowing.

Horses and ponies, which can survive on poor rough grazing, have also been used on some nature reserves. They are selective grazers and may produce very uneven herbage which may favour many different insects and other invertebrates.

I once tried to use a tethered goat to maintain woodland rides. the animal was attached to a cable which allowed it to feed on a strip some 10 m long and 4 m wide. When this had been sufficiently grazed, the goat was moved along the ride to another strip. Goats tend to browse when they have the chance, and the shrubs were bitten down to the ground. By adjusting the time for which each area was grazed, considerable diversity could be developed.

In all these grazing experiments it has appeared that, for successful results from the point of view of conservation, animals should be under the control of those managing the reserves. Otherwise legitimate commercial considerations may ruin an experiment. For instance, we often wish to stop grazing when a substantial amount of food remains. But at the same time it is important that animals used for conservation should be well managed, that disease should be prevented (for instance, sheep become heavily infected with worms if they graze too frequently on the same area) and that nothing should be done to arouse criticism for bad stockmanship among local farmers who we may hope to persuade to go some way along the same road.

Although changes in livestock farming have had the effects which have been described in this chapter, the most important result can be observed in areas outside those in which the animals are raised. We know that it is on arable land that wildlife is at the greatest disadvantage. In Britain about half of the total arable area grows food for livestock, not for man. Out of ten-and-a-half million tonnes of barley grown on 2.4 million ha of land, nine million tonnes go to livestock. Half our home-grown wheat goes the same way. Other parts of the world are also affected, as we import much animal food, grain and soya from America, and fish meal made from anchovies caught in the Pacific Ocean off Peru. Thus intensive livestock farmers may appear to have little direct effect on much of our wildlife, but by using so much of the produce of arable farming they are responsible for many of the changes in land use which have generally had such unfortunate effects.

CHAPTER 5

THE LIVING SOIL

FARMING depends on the soil. The soil may be fertile, growing good crops and supporting healthy animals. It may be infertile, providing the farmer with difficulties if he is to wring any return from the land. Fertile soils may be destroyed, and infertile areas may be improved. The soil consists of mineral material derived from the rocks, various organic and inorganic substances, and a whole series of living organisms – bacteria, fungi, algae, higher plants, protozoa, worms, arthropods and burrowing mammals. These organisms are part of the wildlife about which this book is concerned. Their numbers and distribution are affected by the chemical and physical properties of the soil, and by the way in which the soil is managed by man. They also contribute to the ways in which soils evolve, to the way chemicals circulate within the soil, to the breakdown of some substances and to the ways others are produced, and they are clearly related to the whole problem of fertility. A fertile soil usually contains a healthy and flourishing flora and fauna, though we still do not fully understand all the inter-relationships. Also we recognise that some soil organisms are, from man's point of view, pests, in that they interfere with husbandry and damage or destroy crops.

As pre-agricultural Britain was mostly covered with deciduous forest, its 'natural' soil flora and fauna was, in many parts of the country, that of this habitat. This we find to be very prolific, particularly on mineral soils which are not very acid. It has been estimated that in the top 15 cm of the soil, in one hectare of deciduous forest, we may have nearly 20 tonnes of bacteria and fungi, and two tonnes of earthworms. Slugs, snails, and woodlice amount to nearly three-quarters of a tonne. The protozoa contribute over 300 kg, nematode worms about 40 kg. Insects and mites, though very numerous, with millions of individuals, only weigh in total about 10 kg. These figures are probably maxima, but give the order of magnitude for the different groups.

Forest soils are rich in organic matter, which may amount to over six per cent of its weight. Very little undecomposed material remains for any period on the surface of the ground. This is because of the action of the living organisms in the soil. Earthworms pull fallen leaves into their burrows. Insects and mites chew leaves and break them down into fragments. Fungi develop on the pulverised vegetable matter, and bacteria help to break it down still further. As a result the residues become incorporated throughout the soil. Many complicated reactions take place during the decomposition of the leaves and of dead roots within the soil, so that nutrients are released to be taken up by living plants and recycled. However, the main end product is humus, a black, structureless and amorphous substance which coats the surface of the soil particles and contributes to its stability and structure.

Any organic matter within the soil is sometimes, erroneously, spoken of as humus. This is wrong. It is more accurate to refer to undecomposed organic material as 'raw humus', and restrict the term 'humus' to the black amorphous material whose genesis has just been described. At any time forest soils will contain both raw humus and true humus. Raw humus during its decomposition may actually reduce fertility by removing nitrogen from circulation; this happens when fresh straw is ploughed into arable soil. Undecomposed organic matter accumulates in wet, acid conditions in bogs to produce peat. This, when used in horticulture and added in mineral soils, increases their organic content, and eventually some of the peat may be turned into humus. Compost and fresh manure, when added to the soil, contain little humus, though in time they contribute to its production. True humus slowly releases nitrogen and this stimulates plant growth.

On sandy or acid soils covered with forest the same sort of soil organisms occur, but in smaller numbers than in more fertile areas. The amount of humus is also less. Other natural areas such as salt marshes and moorlands on high ground above the tree line also contain fewer living organisms within their soil. Unfortunately we have few accurate figures giving comprehensive accounts of the different situations, so it is not always possible to evaluate the effects of agricultural practices except on good farmland.

Our knowledge of the biology and importance of different soil organisms is very incomplete. There is still doubt regarding the numbers of different species which are present in a particular soil. Techniques for studying the soil fauna have improved, and have given very different results, each new technique being more efficient and extracting more animals. Thus studies in 1922 found 15 million invertebrates per acre at Rothamsted. Improved extraction methods with soil from the same area in 1936 increased the count to 68 million, of which 45 million were insects. Figures obtained in 1943 were 1,400 million arthropods per acre, including 950 million mites and 280 million collembola. More recent studies have increased the arthropod count to 4,000 million per acre (10,000 million per ha) or even higher. But we still lack knowledge about their importance, and how they affect soil fertility and crop yields. Also there are still many soils which have been neglected so we know little about the plants and animals living within them.

In terms of biomass, the ten tonnes per ha of bacteria found in some soils is greater than that of any other type of organism. Together with the other microflora these inconspicuous organisms make up ninety per cent of the living material in the soil. It is ironical that, when they are thinking of wildlife, most people ignore the microflora and devote all their attention to more conspicuous but much less important organisms. Thus in one hectare of woodland the weight of bacteria in the soil may be ten thousand times as great as that of the birds desporting themselves in the trees above. There are probably several thousand ornithologists, observing the birds, to every microbiologist studying soil bacteria.

One difficulty about bacteria is that individuals are generally very shortlived, and that population numbers fluctuate so rapidly as some die or reproduce by fission many times during the day. Thus total numbers of bacteria per gram of soil may fluctuate between 18 million and 36 million within the period of 24 hours. It would be possible for almost all the bacteria to be wiped out and for the numbers to return to normal within a few days. However, we generally find the most bacteria in fertile soils with high levels of organic matter. Numbers and the speed of reproduction are considerably affected by weather. More bacteria are present when the soil is warm and moist than during winter.

Bacteria are important in cycling nutrients, particularly nitrogen, in the soil. Different species are responsible for different stages in the cycle by which nitrogen gas from the air is taken up, and transformed from ammonia to nitrite and then to nitrate, the form in which it becomes available to growing plants. Other denitrifying bacteria, by returning nitrogen direct to the atmosphere, are ultimately responsible for losses of potential fertiliser by the soil. In general farming practices which give high levels of organic matter in soil ensure that these have high bacterial counts. Some chemicals reduce the number of bacteria, others serve as food and give rise to increased populations. Chemical fertilisers are said to be harmful and organic manures beneficial to bacterial life, but it is probably the amount of food – organic matter – which controls the bacterial numbers, for where organic matter is plentiful substances like ammonium sulphate do not immediately reduce bacterial numbers. But it must be admitted that though it is clear that bacteria are important inhabitants of most soil, we still have much to learn about the ways in which they are affected by modern farming.

Soil fungi are almost as omnipresent, and in similar amounts, as bacteria. They also are important in breaking down organic matter. Fungi mostly consist of thin branching tubes (hyphae, which in bulk produce a network known as the mycelium) which twist everywhere throughout the soil. They also have fruiting bodies, yielding spores which germinate and produce further hyphae. The fruiting bodies may be small and inconspicuous, or, in different species, handsome mushrooms or toadstools. The fungi play a part in attacking and breaking down plant residues and in producing humus. Partly decomposed mycelium itself forms a substantial part of the organic matter in the soil. This is attacked by bacteria, when the nutrients including nitrates are set free and become available to higher plants.

Soil fungi do not always act beneficently. Some cause serious damage to crops. These include the cause of 'Take-all' disease of wheat, *Ophiobolus raminis*, and of Eye-spot, *Cercosporella herpotrichoides*, also affecting the same cereal crop. These only persist when living or dead wheat plant material remains in the soil, and can thus be reduced by suitable crop rotations. There are many other soil fungi which cause economic damage, and others that are beneficial. These include *Dactylaria brachopaga*, whose hyphae produce constricting collar-like outgrowths which have been found to capture and strangle nematode worms which would otherwise attack plants.

The hyphae of many species, including some common toadstools, have an important association with higher plants, including orchids. Particularly in poor moorland soils some trees grow badly or even die unless the hyphae of these fungi establish what is known as a mycorrhizal association with their roots. The mycelium surrounds the roots with a sheath from which some hyphae penetrate the plant and others the soil. The fungus stimulates the decomposition of organic and mineral matter in the soil, and passes the products to the plant, while it obtains sugars and other nutrients in return. This type of symbiotic association is being increasingly recognised as important to many

plants, including some grown as crops. Soils which have been cultivated for years tend to contain fewer fungi of the kind needed to assist
trees, and this may be important where nutrient levels are low.

Other plants found in the soil include algae, which produce slimy
green patches at the surface in mild moist weather, mosses, lichens and
liverworts. All are involved in nutrient cycling and the production of
organic matter. The way in which farming affects the many different
species is clearly an important subject deserving further investigation.

Among the animals, the immense numbers of tiny mites are probably important, for their faecal pellets (a stage in the breakdown of
the plant material upon which they feed) may constitute a substantial
fraction of some soils in coniferous woods. Nematodes are ubiquitous,
mostly free-living, but some parasitic and the cause of serious plant
diseases. Insects include those which feed on plants, and predators
attacking other animals. But earthworms, which exceed in biomass all
other animals combined, appear to be of most importance in their
effects on fertility and soil structure. At any rate, they have been the
most thoroughly studied.

There has been a good deal of confusion regarding the activities of
earthworms, because observers have not always identified the species
involved. In Britain we have 26 species of earthworm, and these differ
both in morphology and in behaviour. Charles Darwin in his book
Vegetable Mould and Earth-Worms published in 1881 accurately recorded his findings, which included the production over the years in
grassland of a surface layer several inches deep of rich stoneless soil
(vegetable mould) from worm casts, and he also demonstrated that
organic matter is introduced into the soil by worms pulling leaves
down into their burrows. Unfortunately he failed to identify the
species, and to point out that surface casts are almost all made by
Allolobophora longa and *A. nocturna*, while only *Lumbricus terrestris*, which
feeds mainly on the surface, pulls leaves down into the soil. The three
species already mentioned all make semi-permanent burrows which
penetrate deeply. The two *Allolobophora* ingest large amounts of soil
from which they extract any nutriment and then come up to the
surface tail-first to pass their faeces (the worm casts). These may
amount to as much as 200 tonnes of soil per ha; 50 tonnes is commonly
voided in old grassland. *Lumbricus terrestris* spends much of the day
quietly resting just below the mouth of its burrow, where it may be
caught before it can retreat by thrushes which seem particularly adept
at this form of hunting. At night the worms emerge, but remain

attached by their tails to the opening of their tunnel. They can draw themselves back inside in a fraction of a second when disturbed. The animals browse on the plants within reach, and also pull down both dead and living leaves on which they continue to feed underground. Tough items like petioles of ash are slowly abraded and consumed over a number of days.

The remaining species of worms are no doubt important, but this is more difficult to demonstrate. Several live fairly near the soil surface, and do not appear to make permanent tunnels. They feed on plant remains and return their faeces to the soil, and are present in sufficient numbers to make a significant contribution to the turnover of nutrients.

Farming affects earthworms, as it no doubt affects the other members of the soil fauna. When woodland soil is cultivated, the number of worms generally decreases. When the soil is put down to permanent grass, populations increase to levels even higher than those found when the ground was covered with trees. Old turf supplies food and shelter, and levels of organic matter may build up in the soil to as high as 12 per cent. The worms and other animals are the main agents responsible.

We have little quantitative information about the effects of farming on forest soils, but the effects of ploughing grassland are better known. The surface-casting worms, *Allolobrophora longa* and *A. nocturna*, appear to be most easily destroyed and are rare in cultivated fields, their populations building up slowly if grass is planted. However, some workers estimate that it may take up to 100 years to regenerate their numbers. Obviously ploughing a field destroys the stratification produced by the surface-casting worms; the vegetable mould is mixed into the mass of the soil.

Lumbricus terrestris seems less easily discouraged by cultivation, and in a garden where farm yard manure or compost is used generously on a vegetable patch, it flourishes. Its popular name, 'the common earthworm', may not be strictly accurate, for the actual numbers of other species may be greater, but any large worm dug up in a well cultivated garden is almost certainly *L. terrestris*. If organic manures, which the worms eat, are not used, worm numbers will be much smaller.

The main changes in farming practice in recent years which might be expected to affect the soil fauna are first the increased frequency with which arable crops are taken from the same area of soil. There are records of as many as 20 successive cereal crops without a break, though some sort of rotation is still general. In those parts of England where few animals are kept the grass ley has almost disappeared. Secondly, a large part of the wheat straw is burned on the stubble. Thirdly, instead of ploughing the land to prepare a seedbed (and to control weeds) minimum cultivation and direct drilling is becoming more prevalent, and will probably increase further as fuel costs make ploughing more expensive.

The effect of continuous cultivation using only chemical fertilisers is to reduce the soil fauna, and also the flora. With ploughing the levels of organic matter eventually fall in most soils to about 2 per cent, and the nitrogen to 0.1 per cent. Continual cultivation does not decrease this level any further; it is probably maintained by the decomposition of the plant roots. Where farm-yard manure is used the organic matter level remains considerably higher, as does the level of soil nitrogen, and the flora and fauna are also more abundant. Organic farmers consider that this is a sign that their soil is much healthier, and that it is likely to produce more nutritious crops.

Many agriculturalists have been worried by the possibility that continuous cropping may damage the soil structure irreversibly. In 1968 and 1969 these misgivings seemed to have been justified. We had unprecedented heavy rain over much of England in July 1968. The fens were submerged for days, the soil everywhere was saturated. The rest of the summer and autumn continued to be wet, and much of the soil remained saturated. In many fields it was impossible to produce a reasonable tilth so as to allow autumn drilling of cereals. Some which were planted were seriously damaged by waterlogging during the winter and early spring, and no crop was obtained. Many people thought that farming would have to adopt a different policy in the future.

The situation was carefully investigated by the Agricultural Advisory Council, who produced a report 'Modern Farming and the Soil' in 1970. They found that most of the damage had been caused by poor drainage; very few soils could not be cultivated successfully, even with these low levels of organic matter, in normal weather if field drains operated successfully. There were a few soils in the midlands which did provide problems, and where special efforts to increase organic matter were desirable. This could be most easily achieved by putting down the land to grass for a few years.

Nevertheless I think that these findings should be treated with caution, and that they should be regularly reviewed if continuous arable cultivations are still used for even longer periods. Soil with only 2 per cent of organic matter (and with few worms, insects, bacteria and fungi) may require no modification, but it is probably more difficult to work in wet weather and there is some tentative evidence that the grain may be less nutritious than that obtained from land containing more organic matter. This may not matter to consumers who have unlimited access to other ingredients of a mixed diet, but could be important when feeding less privileged populations or when the produce is used for animals, who do not have any choice in their diet. Fortunately other, newer, farming techniques may increase the soil fauna as a side effect (p.32), so the worry about the possible (and they are by no means certain) effects of low organic levels may cease to be a problem.

In an earlier chapter I discussed straw burning, and said that it had little effect on wildlife. With regard to the soil fauna this needs some further comment. Surface-living animals are killed at the time of the burn, but soon recover their numbers, so that these are fully restored by the next season. Animals such as earthworms, living within the soil, suffer no ill effects. However, long-term studies showed that after four years of burning before ploughing, different results were obtained. The number of earthworms was reduced by approximately 50 per cent, and most arthropods showed similar population reductions. The explanation appears to be that burning reduced the amount of organic matter, stubble, straw, weeds, which was ploughed into the soil. This caused a slow reduction in the food available for the invertebrates. That this is the true explanation is suggested by other work where soil organic matter was maintained at a high level, in spite of burning; here worm populations remained high.

An increase in organic matter follows minimum cultivation or direct drilling. In this system the straw is generally burned after harvest, or it may be baled and carted away. The field is left for a short period, two or three weeks, to allow shed grain and weeds to germinate. Then the ground is sprayed with a herbicide, usually paraquat (see p.31). This kills all plant growth above ground, and is then adsorbed onto the soil particles and completely immobilised. Such adsorbed paraquat has never been found to have any further phytotoxic properties.

Immediately the spraying is completed the seed may be sown with a special drill directly into the unbroken ground. In some soils this gave greater than usual yields from the outset. In other cases disappointing harvests were obtained, though it has been found that if direct drilling continues for several years yields often improve to levels above those obtained in the same fields after ploughing.

In almost all cases, direct drilled plots have been found to contain higher populations of worms and other invertebrates than do areas which have been ploughed. It is probable that this build-up of the soil population follows the increase in levels of organic matter. Ploughing causes dead roots and other plant residues to be oxidised, while with direct drilling this does not happen, so their organic matter is retained. In some heavy clay soils comparisons have been made of soil structure after several years of direct drilling with soil normally ploughed and cultivated; the first has a much better structure.

It has been observed that in some cases root development starts more rapidly in soil churned up by ploughing than after direct drilling, when compaction may make it more difficult for the roots to penetrate. However where the soil fauna is richest in the directly drilled area the worms work gradually through the ground, and long before harvest the development of crop roots passes that in the ploughed area. There is little doubt that earthworms contribute substantially to fertility in undisturbed heavy clay soils.

Unfortunately, from man's point of view, these desirable animals may sometimes be considered to be pests. Worm casts in grassland areas undoubtedly contribute to increased fertility, but they can also damage both pasture and recreational grassland. When casts are very plentiful, they may cover an appreciable percentage of the turf, and reduce the amount available for grazing. They also produce bare earth in which weed species, less productive for animal feeding, can germinate and invade the turf. Worm casts cannot be tolerated on golf greens or lawn tennis courts. Here they are controlled by poisons which, unfortunately, kill all species, surface casters and others, in just the place where they might do most good to preserve the soil structure.

Even in some arable fields worm casts can be produced in such quantities that the straw is contaminated and harvesting of cereals made difficult. This phenomenon was noted a few years ago in Fryupdale in Yorkshire. Finally *L. terrestris* sometimes pulls parts of growing plants into its burrows. Substantial losses of leeks have been reported when the young leaves are buried and eaten. Although most worms live mainly on dead organic matter, they also nibble living roots in the soil. This form of root pruning is seldom very harmful, but it may occasionally tend to neutralise the other good effects that worms have on the soil and on the crops growing in it.

Many mammals, including mice and voles, foxes, badgers and rabbits dig holes in the ground, but they do not spend much of their time making burrows in the soil. Our only true soil mammal is the mole. Like most of our other indigenous fauna, the mole is primarily a denizen of deciduous woodland. However, it is possible to walk through a wood without realising that the earth is riddled with anastomosing tunnels dug by moles, and that these animals are running up and down the burrows feeding on the worms and insects which fall in. There are few mole-hills in woodland soil, though careful examination will show evidence of repair work where the roofs of the burrows have fallen in. It is not difficult to recognise where burrows occur, particularly when they run across woodland rides, and if a short section of the roof is removed it will probably be repaired with small pellets of freshly dug earth in a few hours. The evidence of the moles' presence is minimal because in woodland they seldom need to make new burrows, and continue to use old ones for many years, generation after generation using the same tunnel with only minor repairs. These supply sufficient space and food for the existing population, which is commonly in the region of two animals per ha. We do not know why it is not greater, for the soil fauna is sufficient to support many times this number without serious depletion. Many young moles, after weaning, migrate out from the area occupied by the mother, and today, when forest covers so small a fraction of the land, they find themselves on farm land.

Man has provided some areas which are particularly suitable for the mole. Old grassland may support a considerably larger population than woodland. The moles make permanent tunnels, but these are not so long-lasting as those in woodland. They may be destroyed by the hooves of grazing animals. In clay soils, with high populations of worms, once a tunnel system has been dug few new hills (made from the spoil of new tunnels) will appear, except perhaps in hard frosty weather when worms dig deeper and new burrows in the warm earth below the frozen surface are needed. In ground liable to flooding moles will move up to dry ground when the area occupied is inundated. The animals are strong swimmers and can escape from flooded ground.

Ploughing destroys the burrow system. Some moles are killed, but most migrate to hedgerows and ditchbanks. As we have already noted (p.27) reinvasion can take place when the crop has been planted. Some damage may be done by invading moles, which will uproot

PLATE 1. The traditional farming landscape. *Above*, Smacam Down, Dorset, in 1948 showing the outline of the 'celtic fields' – old chalk grassland, much of which had not been ploughed for 2,000 years. *Below*, water meadows on the River Itchen in 1934.

PLATE 2. The arable landscape in 1980. Many areas in England are entirely arable with no livestock and no hedges. *Above*, Salisbury Plain, where hedges were never planted. *Below left*, a few trees are all that is left of former hedges in Huntingdonshire; *right*, straw-burning has destroyed many hedges and trees.

PLATE 3. *Above*, a field near Monk's Wood photographed in May 1980. This was arable for several hundred years, and was the site of a Romano–British village; it has been allowed to regenerate naturally since 1961 and is developing into an oak forest. *Below*, grass soon turns into scrub unless it is grazed, as on the right.

PLATE 4. *Above*, grass and clover ley provides good grazing but is of little botanical interest. *Below*, old grass, rich in herbs, being measured for numbers of species, using a point quadrat frame.

PLATE 5. *Above*, Galloway bullocks grazing part of Woodwalton Fen National Reserve as part of its management. *Below*, this grazing produces a diverse herbage and encourages flowers which would otherwise be smothered by coarse grasses.

PLATE 6. *Above*, sheep overgrazing chalk grassland; in spite of this there is floral diversity. *Below*, cattle grazing poor swampy meadowland of considerable botanical interest.

PLATE 7. Ploughing, whether with horses or by tractor, still attracts gulls and other birds which eat the turned-up soil fauna. Modern cereal crops directly drilled do not provide this extra food. Horses are now almost entirely replaced by enormous machines like this maize harvester (*below right*).

PLATE 8. *Above*, forestry, mainly with exotic conifers, is taking over much of our uplands and reducing the area available for grazing by sheep. The pattern of wildlife is also affected. *Below*, 'muirburn' – burning the heather carefully to produce a mosaic of plants of different ages to feed grouse and sheep.

growing plants. The damage is greater than in permanent pasture or woodland as the animals have to construct a whole new burrow system. Studies of moles in arable crops have given the impression that the animals spend most of their time digging. This is only true when no old tunnels can be used. The mole in its natural woodland has a much easier time; some may never dig except to repair their burrows. As was stated in chapter I, very large fields are only invaded round the edges, so if farmers continue to remove hedges and enlarge their fields, the mole may become rarer. As long as we have a substantial amount of woodland, the mole is unlikely to become extinct, as residual populations are likely to survive, unseen, in their natural habitat. The planting of conifers rather than broadleaved trees may make the woods somewhat less suitable, as the soil fauna will be sparser, but contrary to common belief moles do survive quite well in both newly planted and mature pine and spruce forests. The Tentsmuir forest, in Fife in Scotland, is one of our oldest coniferous forests, and it is also well populated with moles.

Moles do no damage in woodland. To the farmer they are a minor pest, reducing the value of pasture and damaging arable crops. They may make some minor contribution, by their burrows, to land drainage, but this seems of little significance. On grassland, as Charles Darwin observed, they bring up stones in their molehills and disrupt the layer of 'vegetable mould'. Moles eat earthworms, and are criticised for doing so, but they seldom seem significantly to reduce worm populations. This is fortunate as earthworms are generally beneficial, improving fertility in most soils.

Many other soil animals and plants are also important in recycling soil nutrients. The soil is indeed living, and in some cases man has adapted the living cycle so that he obtains increased yields of crops, of grass and, ultimately, of grazing animals. He increases the numbers of some forms of wildlife, and he benefits from this increase.

However crops can be grown without the benefit of the living soil, without organic matter, in fact without soil itself. This is by using water containing the necessary nutrients, what is generally known as 'hydroponics'. This has been commercially developed in glasshouses, and in the future may be used for many crops. Actually some field crops, irrigated in sandy soils, are grown by what is very nearly a hydroponic system, as living organisms are very scarce. Some scientists consider these hydroponic techniques will be increasingly successful, and may even replace ordinary farming methods which depend on a healthy, living, soil. But for the present we must rely on cooperating with wildlife, in this case with the natural flora and fauna derived from that which existed seven thousand years ago in the forests of Mesolithic Britain.

CHAPTER 6

HEDGES AND TREES

THE most obvious effect of modern farming, particularly in eastern England, is the disappearance of the hedgerows. From the windows of our farm house near Huntingdon we cannot see a single hedge except for the one around our own garden. Until 1945 tall hawthorn hedges lined both sides of the roads, and surrounded every field. They have all gone.

This is an extreme case, but other areas have suffered almost as severely. On the other hand there are still many luxuriant hedges in the West of England. The situation is obviously different in the various regions of Britain. Unfortunately it is difficult to obtain accurate statistics of hedge removal. Even data regarding the length of hedges present at any one time are of doubtful value. Thus we have had estimates published by one experienced geographer suggesting that, in 1945, there were 2,300,000 km of hedges in Britain, while others quote a much lower figure of 800,000 km. Subsequent investigations suggest that the lower figure is more nearly correct. There have been equally different estimates of hedge removal. In the years leading up to 1970, figures for loss per annum in the whole country ranged from 2,400 km to over 8,000 km. The lowest figures were those of the Ministry of Agriculture, Fisheries and Food, and were based on work done with government grants for hedge removal. Recent studies suggest that, during the period since 1950, the rate has averaged about 8,000 km each year, and that over Britain as a whole nearly 20 per cent of hedges have disappeared. The process continues, though at a reduced rate, partly because, as in my own area, there are no more to remove.

The situation is indicated by the figures in table 1 which is based on results obtained in seven different English counties, and which compares field size and hedge survival in 1945 and 1972. The Cambridgeshire site, in the fens, has no hedges, but fields are divided by ditches which have been filled in in some cases. The field size has more than doubled in the two sites in Eastern England where cropping is mainly arable. Hedgerow removal has also been greatest. Figure 4 is taken from the Huntingdonshire site, and illustrates the situation in 1945 and

TABLE 1 Field size and hedge removal in different parts of Britain (after Westmacott and Worthington, 1974)

County	Cambridge-shire	Huntingdon-shire	Dorset	Somerset	Hereford-shire	York-shire	Warwick-shire
Type of farming	Intensive arable	Extensive arable	Arable and mixed	Dairying	Mixed	General cropping	Livestock dairying and mixed
Average field size 1945, ha	6	8	7	3	65	5	
Average field size 1972, ha	13	18	9	5	7	8	6
Increase in field size 1945–72 %	128	137	16	44	45	26	25
Length of hedge removed, m/ha	0	27	6	15	14	11	9
Length of hedge remaining, m/h	0	43	63	94	148	71	125

FIG. 4. Drawing of the same area in Huntingdonshire in 1945 (*top*) and 1972 (*bottom*). (After Westmacott and Worthington 1974)

1972. At the earlier date many fields were still in grass, at the later all are growing cereals.

In the other areas, where livestock is kept and many fields still grow grass, fewer hedges have been grubbed up and the increase in field size has been more modest. The length of hedge remaining is two to three times as great as in the eastern counties. In general it can be said that most hedges remain in the areas which were best supplied with them at the beginning of the period.

It is widely believed that hedges are comparative newcomers to our countryside, and it is therefore suggested that their removal has little ecological importance. It is true that a great many hedges were planted between 1760 and 1820, when common land was enclosed as the result of Acts of Parliament. However, enclosure acts covered a longer period, the first being passed in 1603 and the last in 1903. Also nearly three quarters of the enclosures were made without this legal process. However, quite a number of hedges are much older, some having certainly existed from Saxon times. These are often parish boundaries, or are associated with special geographical features and patterns of settlement. Most hedges were deliberately planted and managed to make them stockproof, but some of the oldest may have grown up from strips of woodland which were left when assarts (clearances for farming) were made in the forests. Some have clearly developed spontaneously from scrub which has grown up on field boundaries.

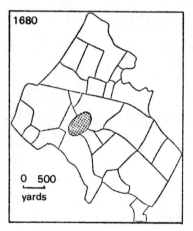 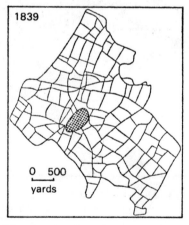

FIG. 5. Increase in hedges, Buckworth, Huntingdonshire, between 1680 (*left*) and 1839 (*right*). (After Pollard, Hooper and Moore 1974)

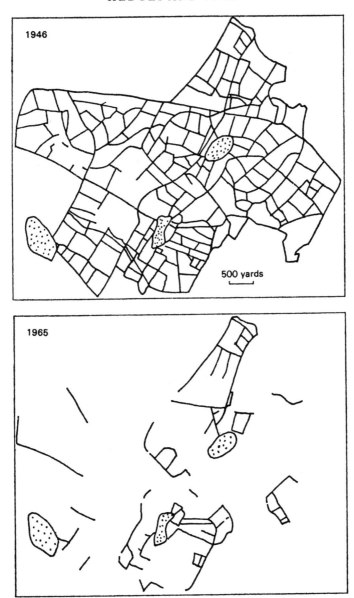

FIG. 6. Loss of hedges in parishes of Barham, Buckworth (see also fig. 5) and Leighton Bromswold, between 1946 and 1965. (After Pollard, Hooper and Moore 1974)

Nevertheless it is true that most hedges are under two hundred years old. This is illustrated by figure 5. In 1680 most of the parish of Buckworth in Huntingdonshire had been enclosed, but only into large fields which were mostly grazed by sheep. The industrial revolution increased the importance of cereals, potatoes and other crops for the urban worker, and fields were subdivided into areas of approximately 8 hectares, suitable for arable cultivation with horses. New hedges were planted, and by 1839 the whole appearance of the landscape had been changed. This same parish is included in the area depicted in figure 6 which shows the situation in 1946, which differs little from that in 1839, and in 1965 by when the majority of hedges had disappeared. The general picture is that, in most parts of Britain, hedges reached their maximum in the middle of the nineteenth century, and almost all remained for a hundred years, until 1950. Removal since that date has been more rapid than planting ever was, even in the most active period after 1760.

There is one difference between hedgeless areas in 1680 and 1980. At the earlier date there were more deciduous woods to support wildlife. It is the areas with fewest hedges in 1980 that have fewest woods, and the fewest native trees.

Farmers remove hedges for many reasons. First they like large fields to use big machines needed for mechanisation and intensive arable production. Tall hedges and hedgerow trees cast their shade and slow down crop development so that some is unripe when the rest is ready to harvest. Although hedges were mostly planted to contain livestock, it is difficult to keep them stockproof, and modern grazing regimes do not fit well into the old field system. Smaller areas are now intensively grazed, and wire or electric fences allow for more flexibility. Hedges are costly to maintain properly. In the old days there was ample surplus labour during the winter months, and farmers had to keep their workers employed. Today labour is more expensive and is generally stretched all the year round. Hedges (and ditches) take up valuable space which could be used to grow crops. The amount of land varies greatly, depending on how the hedge is maintained. If the width that cannot be cultivated is only 2 m (and some old hedges are very much wider), removal of one kilometre of hedge will free a fifth of a hectare, and if the land is worth £5,000 a hectare this is a gain of £1,000. Since 1950 hedgerow removal has probably added 100,000 ha to the land available for agriculture. Hedges are also thought to harbour pests and weeds, which use them as a base to invade the crops.

Until 1972 it was possible for farmers to obtain a grant from the Ministry of Agriculture to contribute to the costs of hedge removal, where this could be shown to add to the efficiency of their farms. The grants only covered part (a third to a quarter) of the actual costs, but conservationists resented the expenditure of public money for what they considered to be the destruction of the countryside. Allegations that farmers removed hedges in order to collect the grant were obviously absurd. In fact most farmers did not take the trouble to apply for grants when removing hedges, saying that the paper work and 'red tape' made this too difficult. Between 1957 and 1969 grants were given for the removal of an average of 1,200 km of hedge per annum. These were the years when most hedges were grubbed up, so less than a fifth of these operations received government support. Grants for this purpose are no longer being awarded.

There are clearly good agricultural reasons why hedges should be grubbed up. Nevertheless most of those who are not practical farmers would like the hedges to remain. They think that they add to the beauty of the countryside. This view has not always been so widely held. The poet John Clare and others, during the period of the enclosures, deplored the loss of the open landscape. Mid-nineteenth century writers also complained about the waste of good land. Thus Chandos Wren Hoskyns in his book 'Talpa or the Chronicles of a Clay Farm' published in 1857 writes as follows: 'It would be a difficult but interesting task to make out a calculation of the economy per acre, of the riddance of these hideous and useless strongholds of roots, weeds, birds and vermin that afflicts the farms of merry England. Unproductive in themselves of anything that is good – for even the timber they contain is rarely so – they are equally an obstruction to the plough that toils for bread, and the eye that wanders for beauty'. The only virtue of hedges this author could find is 'hedge-pheasant shooting – beating the outsides – that pleasant October skirmishing that precedes the coming up of the heavy artillery at Christmas; but is it not rather dearly retained, when land is being cut up for railroads all around us, at two or three hundred pounds the acre, and scarcely a vestige or margin left to enclose for the 'more more' cry of an increasing population'. So over a hundred years ago farmers were concerned about losing land to railways (as now to motorways), with the wish to produce more 'Food from our own Resources' and with the need to grub up hedges. It is surprising that so few were in fact removed for at least ninety years, until after the 1939–45 war.

The aesthetic objection to hedge removal may be subjective. At one time they are 'hideous and useless', at another a 'thing of beauty and a joy for ever'. We do not know how future generations will react. But their importance for wildlife conservation is beyond doubt. When hedges are lost, the quantity and quality of wildlife, both flora and fauna, is greatly impoverished. This, the loss of wildlife, is the main topic with which this book, and this chapter, is concerned.

Unfortunately it is difficult to produce convincing *agricultural* reasons why a farmer should retain his hedges. Only if he keeps livestock which are overwintered outside, and particularly if the farm land is more than 300 m above sea level, is their value as shelter important. In Britain it has been impossible to show that crop benefits from this shelter, except for some delicate horticultural crops. In Denmark and north Germany different results have been obtained, but these are being increasingly criticised. It is suggested that cereals have been found to grow better near barriers in that region because they get more water, resulting from the melting of snow drifts in the lea of the hedges. Hedges may reduce soil erosion by wind or water in some places, but this is seldom significant in Britain. Wind blows of peat soils in the East Anglian fens are serious but these areas never had hedges. Experiments are being made with quick-growing willow hedges, and with rows of barley between vegetable crops, but it seems that these soil blows are more likely to be controlled by good husbandry, and by growing crops which do not leave much bare soil in spring (when three dry days produce conditions allowing the peat to blow even after the wettest winter). Wind erosion sometimes occurs on light sandy soils, and here hedge removal may have exacerbated the problem, but few farmers think replanting would be economically worth while. Some ecologists have suggested that beneficial insects, attacking crop pests, may overwinter in hedges, and that their value may outweigh the damage of pests lurking in the same hedges. There is some substance in the suggestion that bumble bees nesting in hedge bottoms are important pollinators of field beans and other crops, and that in very large fields such insects may seldom reach the middle, but otherwise there is little evidence that beneficial predators from hedges are of any practical significance. On balance pests and disease organisms probably do more damage, as is indicated in a recent leaflet of the government's Agricultural Development and Advisory Service (November 1979) on Barley Yellow Dwarf Virus, which is a serious disease of this cereal crop. The leaflet shows that in 1979 the disease was of little importance in large fields where all hedges had been removed. It was only serious enough to justify spraying with insecticides in those areas of fields near to hedges, particularly those with a tussocky bottom favourable to overwintering the aphid *Rhopalosiphum padi*, the vector of the virus. So the hedges most valuable to other wildlife were just those to pose an agricultural problem.

The only good economic reason which can be advanced for keeping and extending hedges is that they provide cover for game. The shooting rights on some farms are valuable, and outweigh the losses the hedges may produce. But in general we must admit that hedges are valued almost entirely for aesthetic and wildlife conservation reasons. As is shown in chapter 12, many farmers are prepared to retain some hedges and other areas with shrubs and trees because they wish to maintain the beauty of the countryside, and because they are interested in wildlife (and shooting). It is important that conservationists should not try to deceive them, and pretend that hedge retention is a means to economic prosperity.

Conservationists have generally objected to hedgerow removal on the grounds that they provide habitats for wildlife. With the loss of tree cover they often provide the only conditions resembling a woodland or a woodland edge in which our indigenous wildlife, which we have shown to be mainly that naturally occurring in forest, can survive. Thus it has been shown that many birds can exist in substantial numbers in farmed countryside only if hedges provide sites for nests, and that wild flowers commonly survive in hedge bottoms. So without hedges farmland may support little wildlife.

I wish to put forward a somewhat different point of view. Hedges are not only important in providing a habitat for wildlife. Hedges, and hedgerow trees, are themselves important aggregations of wildlife in their own right. Many hedges consist entirely of native shrubs, species seldom seen elsewhere in many parts of Britain. A mature oak tree on a farm is itself the largest chunk of wildlife we are likely to find; it also provides conditions which support hundreds of species of insects and other animals, plants including mosses, lichens and algae. If hedges and trees are removed, this is a direct reduction of wildlife as well as a reduction in habitat for other organisms.

However, the situation is not quite as simple as I have suggested. It may be asked: 'How is a hedge, planted by man, to be valued as wildlife?' If it consists entirely of nursery-grown hawthorn plants, although these are of genuine British provenance, it is surely just a man-made artefact, at least at the time of planting. If it consists of imported stock, or if some exotic species such as *Lonicera nitida* is planted, then this is not British wildlife by any reckoning. An oak tree, even when the acorn which produced it comes from a nearby wood, if deliberately planted by a farmer on his land, may be thought of differently from one which grows up in the same place from an acorn buried by a jay.

I believe that we should generally welcome the planting of in-digenous plants in hedges, woods and shelter belts. Ornithologists are trying to maintain wild populations of rare birds by breeding them in captivity and then releasing them. Entomologists do the same with rare butterflies. It is true that birds and butterflies fly away and give the appearance of being 'wild', though special measures have to be taken to discourage some birds from continuing to depend on their breeders for food. Trees, shrubs and herbs remain on the spot where they are planted, but in our man-made landscape it is surely legitimate to encourage them in situations where natural regeneration would otherwise be slow or even absent.

Fortunately even the most artificial hedge or wood is soon invaded by wild shrubs and herbs. The work of M. D. Hooper on dating hedges is well known. He has shown that the older a hedge, the more species of shrubs it is likely to contain. An observer is instructed to count the number of species of shrub in as many 27 m (30 yard) stretches of a hedge as is possible, and then work out the average. For a pure, newly planted hawthorn hedge the figure would be one. It appears that one new species is added roughly every 100 years, so if the average count is five, the hedge has existed for 500 years, and if it is ten, the hedge will be 1000 years old. This is only a very simplified account; fuller details are given in *Hedges* (Collins, New Naturalist No. 58) by Pollard, Moore and Hooper in their appendix describing 'The hedgerow project for schools'.

The hedgerow dating exercises suggest that new species only de-velop slowly, but invasion has been found to be very much more rapid. A mile (1.6 km) of hawthorn (all *Crataegus monogyna*) hedges was newly planted at Monks Wood in 1962. In 1971 no less than 2,500 seedlings, more than one per metre, were found. The details are given in table 2, which shows that rose (*Rosa canina*) was commonest, then hawthorn with blackthorn (*Prunus spinosa*) third. It is unlikely that many of these first invaders would have survived, but they show that the hedge provided a habitat for indigenous species brought there by natural

TABLE 2 Numbers of shrub seedlings which colonised 1.6 km of
young hawthorn hedges at Monks Wood 1971 (after Pollard 1973)

Rose (*Rosa canina*)	1417
Hawthorn (*Crataegus monogyna*)	920
Blackthorn (*Prunus spinosa*)	161
Ash (*Fraxinus excelsior*)	152
Oak (*Quercus robur*)	51
Field maple (*Acer campestris*)	7
Privet (*Ligustrum vulgare*)	5
Service (*Sorbus torminalis*)	3
Buckthorn (*Rhamnus catharticus*)	2
Dogwood (*Thelycrania sanguinea*)	1

means, some seeds being carried by birds and deposited in their
droppings, others being blown there by the wind. This was also shown
in Hayley Wood, a reserve of the Cambridge and Isle of Ely Naturalists' Trust. Here a pure hawthorn (*C. monogyna*) hedge was planted
in 1971 parallel and adjacent to an ancient, multi-species boundary
hedge. By 1979 the new hedge had been extensively invaded with field
maple, ash, oak, rose and elm, and might have been thought to be
many hundred years old if the dating technique had been used blindly
without examining the total situation.

The colonisation of hedges by non-woody plants has also been
studied. At least 600 plant species have been found in hedgerows, half
of these frequently. Some appear almost as soon as the hedge is
planted; these are generally the common weeds of arable land. Others
take much longer. Thus dogs mercury, a typical woodland plant, is
usually found only in hedges produced from woodland, or in old
hedges, as it seems to take up to ten years to cover one metre from an
existing focus. Primroses, bluebells (*Endymion non-scriptus*) and wood
anemones (*Anemone nemorosa*) are also found mainly in old hedges or in
those which are woodland relics. No species has been found in hedges
and nowhere else, the majority being reasonably common in nearby
areas of woodland. So, as long as any deciduous woods remain, no
species is likely to become extinct no matter how many hedges are
removed. However, without hedges, plant numbers will be much
lower, and fewer people will be able to see many of our most attractive
wild flowers. As conservation is not simply about saving rare species
from extinction, the importance of retaining many hedges is clear.

Without hedges, we should also see far fewer birds in most parts of Britain. As with plants, no species is entirely restricted to hedgerows, though some, such as the lesser whitethroat (*Sylvia curruca*), are seldom found nesting anywhere else. Many woodland birds seem to prefer 'real' woodlands, only moving into the hedges when the woods cannot provide sufficient territories. Nevertheless hedge removal reduces bird numbers and restricts many species to much smaller areas of the countryside.

The way a hedge is managed affects its suitability as a wildlife habitat. During the nineteenth century, when hedges were used to restrain cattle and sheep, they were skilfully laid. When vertical stems were three to five metres tall, a proportion was cut out and the remainder partially severed at ground level and bent over at an angle. They were then fixed to vertical stakes which had been driven in at approximately metre intervals. Some rods of pliable wood might then be woven in to make a stockproof barrier. If enough growing stems were available, these could be omitted. The stems sent out sideshoots which soon produced a strong impassable barrier. For a good many years this only needed to be clipped.

Modern farming has affected the hedges which remain. In some parts, particularly in Leicestershire, Western England and Wales, beautifully laid hedges are still found. In others they are generally neglected. Some are reduced to a few remnants, just a row of unhappy-looking bushes. Others have grown into considerable thickets. With heavy grazing, or where rabbits are common, the undergrowth and lower branches may have been lost. In other areas where the hedge is still a barrier it may have spread untidily outwards covering several metres of precious ground.

Today where hedges are managed, this is usually done mechanically, using cutter-bars or flails. These latter shred and tear the branches and make what looks like a terrible mess, but the shrubs survive and grow again into quite a pleasant looking result. It is then kept in shape with a cutter-bar. The disadvantage of mechanical cutters is that they generally destroy sapling trees before they can grow sufficiently to be easily avoided.

Figure 7 illustrates the different hedge types. N. W. Moore and his team investigated the attractiveness of these to birds; their results are given in table 3. They found that open-field species, the red-legged partridge, skylark and corn bunting, are mainly found on open un-hedged field boundaries. The remnant hedges contain some of these

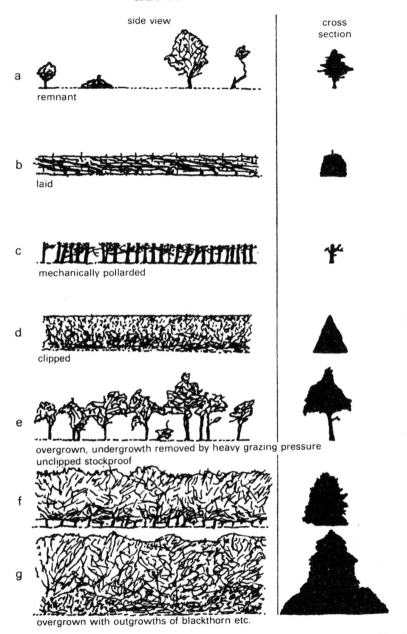

side view

cross
section

a

remnant

b

laid

c

mechanically pollarded

d

clipped

e

overgrown, undergrowth removed by heavy grazing pressure
unclipped stockproof

f

g

overgrown with outgrowths of blackthorn etc.

FIG. 7. Hedge types produced by different management. (After Pollard,
Hooper and Moore 1974)

TABLE 3 Birds in hawthorn hedges under different type of management. (After Moore, Hooper & Davis 1967)

Type of hedge	Length m	Red-legged partridge (Alectoris rufo)	Skylark (Alauda arvensis)	Corn bunting (Emberiza calandra)	Reed bunting (Emberiza schoeniclus)	Whitethroat (Sylvia communis)	Yellow-hammer (Emberiza citrinella)	House sparrow (Passer domesticus)	Song thrush (Turdus philomelus)	Blackbird (Turdus merula)	Linnet (Acanthis cannabina)	Bullfinch (Pyrrhula pyrrhula)	Dunnock (Prunella modularis)
Unhedged boundary	3776	6	4	5	1	2							
a. Remnant	2121	2		4	2	2	2	1	1				
b. Laid	2249	1		1	1	4	1			2	3	3	
c. Mechanically cut	2245		2			2	1	1	2	1	2	1	1
d. Clipped	2565	3					5	6			4	3	4
e. Overgrown (no undergrowth)	2135									2			
f. Unclipped stockproof	2419	1		1	5	6	6	2	6	11	6	2	5
g. Overgrown	1919	1		1	5	6	6	2	6	11	6	2	5

and a few woodland species. The hedges which are laid in the traditional manner are a little richer, but do not compare with the massive overgrown hedges which have five times as many individuals and two and a half times as many bird species.

During the last hundred years bird numbers in hedges, and on farms, must have fluctuated considerably. A hundred years ago when most hedges were well maintained, the bird population must have been fairly low. With the depression when hedges were neglected, and when shrubs encroached on much farmland, bird numbers must have

Pheasant (*Phasianus colchicus*)	Wood pigeon (*Columba palumbus*)	Chaffinch (*Fringilla coelebs*)	Blue tit (*Parus caeruleus*)	Carrion crow (*Corvus corone*)	Robin (*Erithacus rubecula*)	Tree sparrow (*Passer montanus*)	Great tit (*Parus major*)	Goldfinch (*Carduelis carduelis*)	Blackcap (*Sylvia atricapilla*)	Garden warbler (*Sylvia borin*)	Lesser whitethroat (*Sylvia curruca*)	Turtle dove (*Streptopelia turtus*)	Wren (*Troglodytes troglodytes*)	Average number of pairs per 1000 m	Numbers of species
														4.8	5
														6.6	7
														7.1	8
1														6.3	10
	1													10.2	7
	6	3	1	1	1	1								7.0	7
	6	3			1		1							16.3	9
	6	3						1	1	1	3	2	3	37.3	19

increased greatly. Today where neglected hedges are just remnants, birds are scarce. Where neglect has led to vigorous growth numbers are probably high. Where mechanical cutting is practised, numbers are lower than where other forms of management are used. As land is now so valuable few farmers can afford to allow their hedges to grow into mini-forests, the overall result must be a reduction in bird numbers, though where only a few hedges persist the total species count may be maintained.

People are often surprised at the small numbers of birds in the British countryside. Table 3 shows that it is only in the overgrown hedges that we have as many as one pair for 25 m of hedge; laid and cut hedges only have one pair in 150 m. Gardens are often much more populous, with a far higher density of nests in bushes and shrubberies. This is probably because householders supply the birds with so much more food, particularly in hard weather, than can be picked up in the middle of an arable farm.

Small mammals, reptiles and amphibians are commonly found in hedge bottoms and ditches along the side of fields. Most do not stray far out into the open, so when the hedge goes so do the animals. Thus the bank vole (*Clethrionomys glareolus*) is seldom trapped except in and within a couple of metres of a hedge. On the other hand field mice are found all over the largest fields in much the same numbers as may be caught in hedges or woods. I have already described how moles retreat to hedges and ditches when a field is ploughed, and then burrow out for a limited distance as the crop develops and the ground is not disturbed.

Many insects in hedgerows live on host plants which are absent from arable, or even from grassland, and predators live on these insects. Slugs, snails and many other invertebrates are also restricted to undisturbed hedge bottoms. So the number of species which are affected when fields are enlarged and hedges and ditches removed is very considerable indeed.

Hedges are less difficult to remove than large trees, which are often left in the middle of huge arable fields. Unfortunately few are very good specimens; according to Chandos Wren Hoskyns they never were. Some are dead, most are overmature and many, particularly in arable country, are 'stag-headed' with dead branches sticking out of the crown. As late as 1951 the Forestry Commission estimated that there were 73 million hedge and park trees, containing 27.6 million cubic metres of timber, this contributing substantially to the stock of standing timber in the country. But even then replacement was not keeping up with death and destruction. Today the situation is much worse. Dutch elm disease has killed some 17 million trees, and in the west of England this has left no large trees on many farms. The cause of stagheadedness is not fully understood, but may be connected with lowering the water table when fields are better drained. If so this is an important result of good farming practice, which may make replacement of dead and dying trees more difficult.

One reason already mentioned why hedgerow trees are not being replaced is that modern mechanical hedge cutters destroy saplings. Some large landlords, including the Church Commissioners, are trying to persuade their tenants to take extra care and leave young trees in their hedges; an extra premium is paid to recompense for the slower working needed to recognise and avoid saplings which have grown up naturally.

Even when they are not felled hedgerow trees are damaged by some types of cultivation of the fields around them. Deep ploughing close up to the trunk destroys many roots; this is more often fatal with ash than with oak. Tenant farmers are sometimes accused of using this method when their landlords wish trees to be preserved, in the same way as they are said 'accidentally' to burn hedges when straw fires get out of control. I think that such damage is seldom deliberate, but the results are generally unfortunate. On the other hand if minimum cultivation and direct drilling become more prevalent, damage to tree roots will be significantly reduced. However, the tendency is for few new trees to be planted in the middle or even directly adjacent to arable areas, but in small patches in sites where root disturbance is not likely to be serious. In western England in areas where livestock farming is predominant, hedgerow trees may still be grown, as their shade (harmful to arable crops) may be appreciated by cattle. The existing hedgerow trees in eastern England, dying and overmature, are likely to disappear, and with them a substantial fraction of the wildlife which depends upon them. It is perhaps unfortunate that large unhealthy trees, with holes for nesting birds, and rotten timber riddled with insect burrows and fungal hyphae, trees with little value for timber, have the greatest conservation importance. It is obviously good farming and forestry practice to remove such 'eyesores'.

At one time forests played a part in the agricultural system. Cattle grazed among the trees, and pigs rooted for acorns. Fuel and timber was extracted to use in the farmhouse. Most of the forest has now been felled, but coppices and small woods still exist on many farms, and are important havens for wildlife. As land prices increase, those which can be easily cleared and cultivated are still being removed, though, as will be shown in chapter 12, other areas less suitable for farming are being planted. There is increased pressure to grow more trees in this country, as we import more than 90 per cent of the timber and wood pulp used, and world supplies are getting scarcer. The Centre for Agricultural Strategy of the University of Reading suggests in a report published in 1980 that we could double the area under trees by the end of this century. Most planting would be of softwoods in the uplands, and little on good agricultural land. As already indicated (p.46), it may be possible to afforest much rough grazing while at the same time improving the remaining areas, and thus produce both timber and sheep with improved efficiency. Such developments will obviously have major effects on wildlife, but they are not our concern in this book.

CHAPTER 7

PONDS, RIVERS AND MARSHES

WATER supports many different types of wildlife. The existence of a pond, a river or a marsh means that many plants and animals which are absent from other farms may be found, living on the farm in, on or around the water. They will be affected by the way the farm is organised, and may in turn have their own effects on the farm itself.

There are still many small ponds on farms in all parts of the country, even though the number is decreasing rapidly. The majority were created to provide water for livestock to drink. Others were dug to obtain clay. Thus in our own garden we have a large pond which produced the material used to fill the spaces between the beams when our house was built in the seventeenth century. Some are marl pits, from which clay was extracted to spread on peat or sandy ground to improve the structure of the soil. The banks of these ponds were generally graded so they could be used by cattle or sheep. Village ponds were used as water supplies, often for drinking, for washing farm vehicles, and to put out fires which were common with thatched houses. Villagers allowed their ducks to swim on the ponds, where they obtained free food – from the wildlife.

Cattle ponds are of many types. Some are simply holes in the clay soil, and in regions with a high water table they remain well-filled except in serious droughts. Many have a layer of puddled clay to make them water-tight. On the downs there are the so-called dew ponds, shallow depressions lined with puddled clay. In the open country they often give the illusion that they are on top of a ridge, but in fact they do drain the rain from quite large areas and do not rely on the deposition of dew to fill them.

Farm ponds were dug in the sites where, at the time of construction, they would be most useful. Many are located where several fields meet, so that animals grazing in any of them had access to the water. In some cases such ponds are star-shaped (figure 8), with an arm of shallow

87

FIG. 8. A star-shaped pond. (After J.Dyson 1976)

water penetrating each field, and with the middle deep enough to discourage animals from crossing into other fields.

Newly dug ponds were quickly colonised by plants and animals. Birds such as duck soon discovered them, though they provided little food until colonisation by plants and insects took place. The ducks brought seeds and pieces of weed on their feet. Insects such as pond skaters and beetles flew there. Frogs and water voles found their way overland. It is often surprising how many organisms appear, and how quickly. A well-used pond had few reeds or willow trees, as these could not compete with trampling hooves of cattle coming down to drink, though some with high banks on part of the perimeter supported trees and other vegetation.

As mains water was brought to the farms it was piped to the fields and the animals drank from troughs. This ensured a constant supply, and diseases carried by dirty water were avoided. So the ponds soon lost their original function. With no further trampling willow and alder trees quickly developed, as did reeds and other water plants.

Nesting birds were not disturbed. The value for wildlife increased as a result of this development.

Unfortunately, from the point of view of conservation, farmers often fill in ponds of no agricultural importance. Some have left them to attract duck, but these are the minority. Those ponds near villages are thought to encourage the breeding of biting gnats and other insects. Garbage and refuse, old bedsteads, and even derelict motor vehicles, found their way into the holes when they were not filled with soil prior to cultivation. Where hedges were removed many ponds were left in the middle of large arable fields, making cultivation difficult and encouraging rabbits which burrowed in their banks and grazed on the growing corn, so it was good farming practice to destroy them.

It is difficult to give an accurate picture of the effect of pond destruction on wildlife. It is blamed, probably rightly, for the fall in the number of frogs and dragonflies in many parts of England. However, it is difficult to be certain that pesticides and fertilisers used in arable fields may not have been partly to blame, a view supported by the disappearance of tadpoles from many of the remaining ponds. Duck and moorhens (*Gallinula chloropus*) are not found unless water remains, and moorhens, which are strongly territorial during the breeding season, are commonest when there are many small pools each serving as a territory for a pair. Where there are fewer but larger water bodies, e.g. irrigation reservoirs, moorhens seem to establish larger territories with smaller total populations. Unfortunately there has been no proper nationwide survey of ponds and of their disappearance, so we have little accurate information on how the flora and fauna have been affected.

There has recently been an attempt to rehabilitate village ponds to counteract the loss of those on the farm. Unfortunately many village ponds have become refuse dumps, while others have been 'suburbanised' with paved banks, goldfish and few indigenous plants. The 'Save the Village Pond' campaign has organised teams of voluntary workers to remove the old iron, clear out mud and decaying vegetable matter, but at the same time to encourage natural vegetation and indigenous fish, amphibians and insects. It has also been found that the increasing number of small garden ponds, if these are not managed too intensely, are being invaded by frogs and toads, which seem to breed there with such success that the loss of rural ponds may be to some extent offset, at any rate as far as these species are concerned. Nevertheless, there is scope for saving the remaining ponds on farms where these do not seriously hinder other work, and possibly for creating new ponds on local nature reserves and other land which is not primarily being used for agriculture.

Rivers and streams flow through, or along the boundaries, of many farms. The nine Water Authorities in England and Wales are responsible for the waters in their areas. Nothing may be deliberately discharged into the rivers or streams without their consent, which is withheld if the discharges are likely to reduce the water quality. In recent years there has been a great improvement in the cleanliness of British rivers, particularly in the rural areas, i.e. the areas where the water flows through farmland. There are still highly polluted stretches of river in industrial towns, though even here some improvement has taken place. The average level of organic matter and of toxic chemicals has, in general, been reduced, and the numbers of incidents when poisonous discharges have produced serious fish kills and other evi-

FIG. 9. Nitrate levels in two British rivers. (After Mellanby 1980)

dence of gross, though temporary, pollution have decreased. The Water Authorities are making real progress.

However, things are not entirely satisfactory, particularly with regard to the use of rivers as sources of domestic drinking water. The level of soluble nitrate in many of our rivers has increased, doubling or even trebling since 1945 (figure 9). The average level of nitrogen is still normally below 10 milligrammes per litre (which is the same as parts per million, ppm), but occasionally this rises to 30 ppm. Even this is quite safe when drunk by adult humans and animals, but could be dangerous to babies and young stock, because the nitrate is transformed to nitrite by micro-organisms in the gut of young animals. This nitrite combines with the red pigment in the blood to form methaemoglobin, which does not carry oxygen from the lungs to the tissues. So far in Britain no serious cases of methaemoglobinanaemia from drinking tap water drawn from rivers have been reported, though a few have occurred when well water (probably contaminated by livestock excrement) has been used in babies' bottles. On occasion Water Authorities have stopped drawing water from rivers for a few days when nitrogen levels have been high. As the levels of nitrate are still rising, there is natural concern that at some future date the situation might be much worse; it is difficult and costly to remove nitrates.

The main reason for the rise of nitrogen in our rivers is run-off from farm land. Nitrogen levels are rising parallel with the increased use of chemical fertilisers. The situation is not a simple one. Of the nitrogen applied to farm land, about half is taken up by arable crops, and grass may use up to 80 per cent. The rest is lost in various different ways, the most important of which is drainage. It should be realised that soluble nitrates may be removed by heavy rain from both farmyard manure and chemical fertilisers. When in the past much farmyard manure was applied in winter, the greater part of the nitrogen was lost before spring, and was therefore not available for the crops when they started to grow. Over a hundred years ago this was shown in experiments on the famous Broadbalk field at Rothamsted, when drainage water from plots with various fertilisers was analysed. There may even be raised levels of soluble nitrogen in drainage water from fields growing legumes which have fixed atmospheric nitrogen by means of their root nodules. Were all our arable farms run organically, and if they all had optimum dressing of farmyard manure, the run-off problem would be little less serious even though nitrogen release from natural manure may be somewhat slower than from chemical dressings.

It is obviously to the farmer's disadvantage for expensive nitrogen to be lost in drainage water. Losses are greatest in winter, when there are no actively-growing crops to take up the chemicals. Dressings of fertilisers applied during the spring and summer are split, so that only part of the total used is present in the soil at any time, and so leaching is reduced. Slow release compounds are used; here the soluble (and leachable) nitrate is liberated gradually, so that it can be taken up by the plants as it becomes available. These and other techniques are probably going some way to reduce nitrate losses, but not sufficiently to neutralise the effects of increased fertiliser use.

The Water Authorities have no control over farming in their catchment areas except when wastes are directly discharged into the rivers. The run-off water is not put through sewers, which the Authorities supervise, but through a multitude of field drains and even by natural seepage through river and ditch banks. Were the situation to deteriorate it might be necessary for legislation to be introduced to control the amount of fertiliser, but so far this course of action has not been contemplated. The rising costs of chemicals may be the most important reason why greater attempts will be made to reduce losses, and so to prevent what is in effect the pollution of our rivers.

From the point of view of wildlife, the effect of the rise in nitrate levels in our rivers has been negligible. However, more serious trouble has occurred in stagnant drainage ditches and in many streams, where dense algal growth has occurred with reduction of other plants and animals. More serious effects of eutrophication have been noticed in lakes and reservoirs, and in areas such as the Norfolk Broads. Here the nitrate from farmland has probably played only a small contributory role. Parallel with the rise in nitrates there has been an even more spectacular rise in phosphate. This is almost certainly from treated sewage effluent, and comes mainly from the breakdown of detergents used in homes and factories. Phosphate fertilisers are used on farms, but little of this phosphate appears to be lost in the run-off. Were the rivers very low in nitrate, the phosphate would have little effect, but even thirty years ago the nitrate levels then existing would have been high enough, with added phosphate, to give rise to symptoms of eutrophication. So farming methods of using fertilisers probably do not themselves do any direct harm to wildlife to any great extent. The main risk is to humans, and animals using the water.

There is a real risk to fish and invertebrates in rivers from insecticides used in fields on the banks, and to aquatic plants from herbicides. This subject is dealt with in more detail in chapter 9 (p. 105).

Drainage ditches run along the boundary hedges of many fields, and tile drains in the fields themselves discharge into these ditches. In most of Britain the ditches are normally dry, only running intermittently in heavy rain. They therefore support little aquatic life, though plants requiring damp conditions may grow in them. In flat parts of the country, particularly fenland in East Anglia, the ditches are normally filled with water. Most situated between fields are quite narrow, but others which bear names like 'Sixteen Foot' or 'Forty Foot' are wide and may be navigable. They may contain many coarse fish and a rich invertebrate fauna.

All ditch banks provide some cover, used by birds, small mammals and other wild animals. They often have a rich flora. Unfortunately it is the overgrown and neglected ditch which has the greatest value for wildlife. Such overgrown ditches are easily blocked and so do not carry surplus rain away efficiently. Farmers are responsible for keeping field ditches clear, and in doing so they may destroy, and discourage, wildlife. Some are prepared to be less drastic in their clearance than others, to leave interesting plants and uncropped areas, but we have to accept that regular cutting of the vegetation and cleaning out of channels is necessary.

Larger drains are the responsibility of Internal Drainage Boards. I have served for some years on the Sawtry Internal Drainage Board, which is responsible for an area of farmland in the West of Huntingdonshire, and for Woodwalton Fen National Nature Reserve. The other board members are drawn from the farmers in the area. We meet at intervals and prepare plans to ensure that farmland is not seriously flooded. We keep the drains clear, and install pumps to transfer water to major drains where it becomes the responsibility of the Middle Level Commissioners and, ultimately, the Anglian Water Authorities. These Commissioners have a major pumping station which ensures that the surplus water is speedily discharged into the tidal Ouse and via the Wash to the North Sea.

Keeping drainage channels clear and working efficiently is not an easy task. Most of the clearing has to be done by contract labour, and the contractors may put speed before aesthetics and wildlife. Once again the most efficient drainage ditch is the best groomed, so the tendency is to remove all the vegetation on the banks, including trees which prevent the easy access of bulldozers and other earth-working machinery. A well maintained ditch has sides neatly plastered with clay dug from the channel, with no vegetation on the banks, and no plants growing in the water. Perhaps fortunately from the point of view of wildlife it does not long remain in this condition, wild vegetation soon develops, and animals return. As clearing is expensive, ditches are often left for up to ten years during which period their conservation importance gradually increases, to be destroyed again next time round.

My area has been intensively farmed for more than a hundred years, so the work of the Drainage Board consists mainly of maintaining the *status quo*. We are unusual in having a major wetland nature reserve in our area, and here we reverse our management practice and try to keep it over-supplied with water, a condition which existed previously in the undrained fens. This is actually useful to the farmers, as the reserve can act as a wash, taking surplus water in times of flood when it is impossible to get all the water away to the main pumps and the sea. Many other parts of the country have substantial unimproved areas, which the IDBs, and the Water Authorities, may wish to drain, to increase agricultural productivity. This brings them into conflict with those concerned primarily with wildlife conservation.

The main reason for this conflict is that we have so few wetlands left. We know that most of lowland Britain was deciduous forest seven thousand years ago, but there were also substantial areas of marsh and fen, amounting to hundreds of thousands of hectares. These wetlands have been mostly drained and have become good farmland. This process is still going on.

In the fenlands of East Anglia there was little drainage before the Roman occupation. Then ditches were dug, and water was removed from quite large areas of peat and siltland. Much of this was probably inundated and lost during the troubled years after the Roman legions departed, but piecemeal drainage of small areas continued during the medieval period, with occasional more ambitious projects, such as that of Bishop Morton who, in the fifteenth century, caused a great drain, 'Morton's leam', to be cut to take water from the uplands (that is land a few metres above sea level) towards the North Sea. But it was in the seventeenth century when really large-scale drainage was carried out, by the Dutch engineer Cornelius Vermuyden on behalf of the Earl of Bedford and his associates, the so-called Gentlemen Adventurers. The Old and New Bedford rivers date from that period. Great areas of fenland were drained, but the results were not always satisfactory as banks broke at times of flood and inundation occurred. Also insufficient allowance had been made for the shrinkage and oxidisation of peat soils under cultivation, a process which continues so that many drained areas are now well below sea level, and water has to be pumped out into drains which run between elevated banks at levels considerably higher than the cultivated fields.

The last major feat of drainage was the destruction of Whittlesea mere by John Lawrence in 1851. This lake, which in times of flood covered well over a thousand hectares, was surrounded by great areas of reedbeds abounding in wildfowl. After it had gone the remaining Huntingdonshire fens were quickly drained, with the exception of the five hundred or so acres of Woodwalton Fen, which was saved by Charles Rothschild and is now a National Nature Reserve leased by the Society for the Promotion of Nature Conservation to the Nature Conservancy Council.

Much of the drained fens were, in the eighteenth and nineteenth centuries, used for grazing livestock. Unimproved areas of fen, where they remained, were also brought into the system. These fens were generally under water for most of the winter, but in summer, with low rainfall and high evaporation, they dried out so that cattle could be admitted to feed on the natural vegetation. This probably increased the botanical diversity, and encouraged many wild flowers to blossom, as in the recent experiments at Woodwalton fen (p.54). The ditches, even in arable areas, were not managed very rigorously, so waterlilies and other aquatic plants were common. Dragonflies and other insects were numerous. The human population was too low to pollute the water seriously, and little fertiliser or manure was used so the waters were not eutrophic. Although the main fenland areas had lost much of their native vegetation and most of the animals dependent on it, the rich flora and fauna of the dykes meant that the ecological interest of drained fenland was by no means destroyed, until intensive arable farming became widespread in the period since 1945.

Today little drained fen is under grass; the land is too valuable. Wheat, barley, sugar beet, carrots, onions, chicory are commonly grown, and give among the highest yields in the country thanks to the liberal use of fertilisers and pesticides. There are a few patches of trees, mainly around villages. Some willows grow along watercourses, when they have not been destroyed by drainage workers. The banks of the channels are mostly kept cut, the water has little of the rich vegetation commonly found a hundred years ago for this is often controlled by herbicides. All forms of wildlife are at a low ebb. But the standard of farming is excellent, and is likely to continue to be productive and profitable in the future. The silt soils are being steadily improved. There is concern at the disappearance of the peat, but even when it is exhausted the underlying clay, though more difficult to cultivate, will continue to give high yields. Wildlife has poor future prospects, except on sufferance in odd areas of little agricultural value, and in the few nature reserves established mainly by voluntary bodies like county nautralists' trusts. There is good reason for conservationists opposing further draining of the remaining wetlands. Unfortunately for wildlife, these often have such a high agricultural potential that it is difficult to prevent their development. However, there have been some successes for conservation, and at least a few characteristic sites may be preserved for the future.

One area where there is obvious conflict between farming and conservation is the Ribble Estuary in Lancashire. This is given the highest rating, Grade 1*, in the Nature Conservation Review, published by the Nature Conservancy Council in 1976. This means that it is 'of international or national (Great Britain) importance. . . the safeguarding of which is considered essential if there is to be an adequate basis for nature conservation in Britain, in terms of a balanced representation of ecosystems, and including the most important examples of wildlife or habitat'. The mudflats cover some 5,200 ha, and support large populations of wildfowl and waders. It is a main centre for pink-footed geese (*Anser brachyrhynchus*), with some 6,000 birds. The total wader population is around 100,000. The 2,800 ha of marsh are at present partly used as rough grazing, which probably increases their attractiveness to birds. The marshes are recognised internationally for their value to wildfowl, which include shelduck (*Tadora tadora*), redshank (*Triga totanus*), oystercatchers (*Haematopus ostralagus*) skylark (*Alauda arvensis*), blackheaded gull (*Larus ridibundus*) and common tern (*Sterna hirundo*). It is also the southernmost breeding haunt of the dunlin (*Calidris alpina*) in Britain today. The marshes are also of considerable botanical importance, and also support many invertebrates. The area was a priority candidate for acquisition as a National Nature Reserve as soon as funds were available.

Unfortunately before it could be acquired by the Nature Conservancy Council, or the Royal Society for the Protection of Birds which was also anxious to preserve the site, it was bought in order to be drained and intensively farmed. There is no doubt that the land has great potential, and that if it were reclaimed much of the soil would be of the highest agricultural quality. Drainage might not affect the mud flats directly, though many of the birds would not have remained adjacent to a busy, highly mechanised farm. The value of the marshes for wildlife would have been totally destroyed.

It is somewhat ironical that, though the land had been bought with foreign capital, grants paid for by the British taxpayer would have been available towards the cost of drainage. Fortunately at the eleventh hour the government acted and the Nature Conservancy Council was permitted to buy the marshes and designate them as a National Nature Reserve. Pleasure at this happy outcome was somewhat marred by the fact that the taxpayer had to fork out many hundreds of thousands of pounds more than would have been paid if our government had acted with a little more speed. It should be noted that in this instance the case for conservation was unanswerable, as the site was already scheduled as of outstanding importance.

In other instances the conservation case has been more difficult to establish. Thus, as part of the flood protection measures in Norfolk and Suffolk, where a surge from the North Sea caused devastation and loss of life in 1953, it is suggested that a barrier be placed near the mouth of the River Yare. This would normally be open, and would not then interfere with river levels, tidal penetration and other conditions in the river valley. The barrier would only be closed when an unusually high tide was forecast; this would probably be less than ten times (and for not more than two or three tides each time) a year. This would not have any major effect on the river valley and the grazing land, nor on the Norfolk Broads area further up the river. However, one major justification of the barrier is that it would make it easier to drain some 15,000 ha of marshland which is at present providing poor grazing, and turn it into productive arable land. The increased cash value of the crops is calculated as sufficient to offset the cost of building and operating the barrier.

When the barrier scheme was being prepared, the engineers were able to say that it would not directly affect any sites of recognised conservation importance. There were no National Nature Reserves or even Sites of Special Scientific Interest involved. There is some feeling that the belated discovery of the importance of the marshes to wildlife, and the strong opposition to the barrier by the whole conservation movement, is misconceived. There are those who suggest that conservationists are opposed to all change, and that they will make wildlife an excuse for this opposition, even when plants and animals are not seriously at risk.

The case of the Yare valley marshes is a complicated one. Interest in the past has been concentrated on the more restricted area of the Norfolk Broads where there are many reserves of recognised international importance. Unfortunately these have suffered considerably in the last thirty or so years, with increased pressure from visitors and pollution from human activities and, ironically, from the excretion of gulls roosting in increasing numbers on the surface of some of the broads. This deterioration has increased interest in the surrounding land, including the Yare marshes. Although most of the land there is covered with what appears to be rather dull and monotonous grass of poor quality, many wild flowers do exist. However, the major interest is in the dykes which divide the fields. These contain a rich flora and fauna, with rare species like the sharp-leaved pondweed (*Potomogeton acutifolius*) and the larva of the dragonfly the Norfolk aeshna (*Aeshna isoceles*). This insect is so rare that it is included in the list of species which is likely to receive protection by Act of Parliament.

The trouble about the marshes is that they cover such a large area. They are also, at least at first sight, somewhat uniform. Particular areas of outstanding importance that are obvious candidates for protection have not been identified. It has been difficult to make a convincing case for preventing drainage of any area which is only a small fraction of the whole. But if the entire area were changed, this would be most serious, and the first proposals for the barrier envisaged conversion of the whole to arable. With this, some of the drainage ditches would remain, but they would be rendered of little value for the present type of wildlife, being deepened and canalised. It is also possible that if the fields between the ditches were intensively farmed, fertilisers and other chemicals might leach out into the ditch and eutrophicate the water, totally altering its vegetation and its animal life.

Even without a barrier these marshes are in danger. Substantial areas have already been drained by farmers who consider that the risk of an occasional flood is worth taking. Some farmers have not even bothered to apply for a government grant towards their work – as in the case of hedgerow removal they have preferred to get on with the job and to avoid the complication of dealing with a government department. I believe myself that the marshes in this area, and in many other parts of Britain, are in great danger. When the government

PLATE 9. *Left*, this ancient oak tree at New Bells Farm, Houghley, Suffolk, is an enormous chunk of wildlife in its own right; it also supports hundreds of species of insects, epiphytic plants, birds and other life. *Centre*, hedge-laying to produce a stock-proof field barrier and, incidentally, to provide a habitat for many birds. *Below*, old trees and uncut vegetation along the river Stour support a rich fauna.

PLATE 10. *Above*, unimproved marshes in Dumfriesshire are a haven for many species of birds. *Below*, reclaimed saltmarshes are valuable to the farmer, but have lost their myriads of wildfowl.

PLATE 11. *Right*, an area in Lincolnshire of great diversity – trees, dykes, uncultivated patches and arable fields, where a diverse flora and fauna flourishes. *Centre*, the river Ouse, with grazing which floods in winter up to the level of the high banks. This is moderately rich in wildlife. *Below*, drained meadows in Sussex. The grass has been 'improved' and provides good grazing but few flowers. The water contains few plants and animals as it is contaminated by agricultural chemicals.

PLATE 12. *Above*, dyke clearance temporarily destroys the vegetation but (*below*) plants soon re-established themselves and the fauna returns also.

PLATE 13. *Above*, this village pond at Tissington, Derbyshire, contains a wide range of wildlife in addition to being picturesque and supporting domestic duck. *Below left*, this irrigation reservoir also encourages wildfowl in addition to helping to improve crops; *right*, another haven for wildlife – a farm pond.

PLATE 14. *Above*, thatched buildings may be fire hazards, but birds love them. *Below*, the old farmyard was a messy place, but this encouraged flies which supported bats and insectivorous birds.

PLATE 15. *Above*, grass for silage produces valuable, high protein animal food, but the meadows support little wildlife. *Below*, zero grazing of Jersey cattle. These animals do not feed in the fields, so they contribute little to wildlife: cowpats are a valuable habitat for many insects and some fungi.

PLATE 16. Agri-chemicals are necessary to modern agriculture. *Right*, spreading solid fertilizer on grassland. *Centre*, nitrogen injection is an efficient method and causes minimum loss outside the crop. *Below*, modern insecticides protect many farm crops today.

agreed not to give grants to farmers in other areas of wildlife importance, this was, rightly, welcomed as a victory for conservation. But
without a more positive policy, including the setting up of far more
nature reserves on wetland, the future is far from rosy.

Fortunately there has been some increase in wetland habitats,
which goes a little way to compensate for the losses from agriculture.
During this century gravel workings have flooded, and produced
many new lakes, particularly in river valleys in Eastern England and
the Midlands. Modern farming has played no part in this process; in
fact the new lakes have mostly been dug on what was previously good
farm land. Some of the licences to extract gravel have stipulated that
the land be restored to agriculture, and this has generally been successful. However these new areas of water are often more valuable, in cash
terms, than the farmland they have destroyed. Many are used for
recreation as marinas for boats using adjacent rivers, for water skiing,
or for fishing. Where fishing is predominant, wildlife also flourishes.
Fortunately quite a number of these lakes, which quickly become
surrounded by reedbeds and willow thickets, are now designated as
nature reserves. They have obviously encouraged wildlife. This is most
spectacularly illustrated by the increase in numbers of the great crested
grebe (*Podiceps cristatus*). In 1860 it was feared that this species was on
the road to extinction, as only 42 pairs were counted in Britain. We do
not fully understand why it started to increase soon after that date; the
earlier fall in numbers was probably caused by human persecution.
The increase since 1900 is clearly linked with the creation of new
habitats, though this is probably not the whole story. The climate may
have become more favourable. Today most suitable areas of water are
occupied by grebes, and there are probably more than 4,000 adults
living in the British Isles. Invertebrates, fish and plants have also
spread in the same areas of water. Fortunately there is little evidence of
agricultural pollution in gravel pits. Some, unfortunately, have been
used to dump refuse, sometimes successfully covered with soil to go
back to farming. Some refuse has been illegally tipped into others and
has damaged wildlife.

The Water Authorities are continuing to create large reservoirs,
some, like Grafham Water and Rutland Water, in the middle of
farmland in southern England. There has been much opposition to
these reservoirs, from farmers who deplore the loss of good agricultural
land, and from some of the amenity societies on similar grounds. On
balance, wildlife has almost certainly gained. The farmland lost in the

two cases mentioned had no great conservation value, and the reservoirs have been managed so as to satisfy as many purposes as possible. These include recreation, sailing and fishing. But substantial areas have also been set aside as nature reserves, managed by local naturalists' trusts, and these have been very successful. The immediate interest has been in the immense number of birds which visit the reserves in winter, but attention is also being paid to the other fauna and the flora, and in time these artificial habitats will do much to compensate for the loss of other lakes like Whittlesea mere, even if the two water bodies were so very dissimilar in their origin.

Other bodies of fresh water have been created by farmers, to use for irrigation. The statutory Water Authorities control all abstraction from rivers, and are unable to allow all those farmers, particularly in the drier parts of Eastern England, to abstract anything like as much as they would wish during the summer, for unfortunately the river flow is least when the need to irrigate is greatest. This is partly got over by creating reservoirs to store the water, which is collected during winter where there is a surplus. Some of these are strictly utilitarian, rectangular troughs lined with puddled clay, or, more recently, with polythene. Even these are not entirely useless to wildlife, they provide water for drinking, duck are difficult to discourage, and some plants and invertebrates find a home. However, many farmers landscape their ponds, encourage suitable fringe vegetation and provide refuges for duck and other wildfowl. In a few cases quite large lakes have been made providing splendid facilities for immense numbers of wildfowl during the winter. This is one of the few ways in which government grants to farmers, which pay a part of such developments where they are ostensibly for irrigation, actually contributes to wildlife conservation.

CHAPTER 8

FARM HOUSES, YARDS, GARDENS AND BARNS

MANY farm houses in England are ancient buildings, several hundred years old, with thatched roofs and beams even older than the houses themselves. Nearby there is a range of farm buildings of equal antiquity, and a yard containing heaps of ripening farmyard manure with chickens scratching around for fallen grain and insect food. The animals drink from a muddy pond, and a few ducks swim on its surface. Traditionally good farmers are poor gardeners so the house is surrounded with trees, shrubs and unkempt grass. There is often a neglected orchard of old, gnarled fruit trees, encrusted with lichens and full of holes for nesting birds. The whole complex is a veritable oasis in the well cultivated farm, with arable fields neat and weed-free and grass mown or grazed so that it is equally tidy. The house, the outbuildings and the garden offer shelter and nesting sites for a substantial population of birds which feed on the farmland. Insects and wildflowers also abound near the farmhouse, even if they are totally excluded from the acres of arable crops.

However, this idyllic picture is becoming less and less common. Many old farm houses have been pulled down, others have been sold with a few acres to city dwellers, who have retained the facade and renovated the interior. Farmers often live in modern houses which might equally well have been erected in outer suburbia. Instead of stables for the horses, they have garages for their cars and tractors. The gardens are well maintained, possibly because the heavy work may be done by workers charged to the farm account. The outbuildings are modern structures of utilitarian design, and if animals are kept their accommodation is as clinically uncontaminated as a bacteriological laboratory. The farmer and his family may be more comfortable than his ancestors, and the livestock may be healthier, but it is the wildlife that has to pay the price.

Unfortunately there have been no systematic studies of the wildlife of farm buildings and their surroundings, but Bruce Campbell has

101

described the birds which live around his house, garden, orchard, walled garden, tree belt and rough paddocks, some three hectares in the middle of arable and well-grazed farmland. The house and out-buildings are of Cotswold stone roofed with Stonesfield slate. In the best years he estimates that there were 170 breeding pairs of birds. In every season he had pheasants, moorhens, jackdaws (*Corvus monedula*), blue tits, mistle thrushes (*Turdus miscivorus*), song thrushes, blackbirds, robins, dunnocks, starlings, greenfinches (*Carduelis chloris*), chaffinches and house sparrows breeding. Usually woodpigeons, great tits and stock doves (*Columba oenas*) also nested. A great many other species bred in different years, their numbers sometimes following national trends and at other times appearing or disappearing without apparent cause. Thus the barn owl (*Tyto alba*) bred in most years before 1962 but has not done so since. The spotted flycatcher (*Muscicapa striata*) is the commonest summer visitor with five pairs in good years. The cuckoo (*Cuculus canorus*) has laid in dunnock's nests. Over the last 30 years 48 species of birds have nested in these grounds. Although this was not exactly a typical old farm house, the bird populations must have been similar in and around many farms before they were 'improved' by their new owners.

Old farm and other houses usually support numerous animals in addition to birds. Many insects and spiders inhabit human dwellings, generally unnoticed, but sometimes recognised as damaging the fabric or the inhabitants. Norman Hickin has, in his *Household Insect Pests*, described the many arthropods which occur in old houses, including silverfish *Lepisma saccharina*, various cockroaches, flies, fleas and wood-lice. Many species of beetle can usually be found eating our food, our carpets, or the very structure of the house itself. Thus beams and other timber may be riddled with beetle larvae, particularly the common furniture beetle (*Anobium punctatum*), which is present in almost every old house where the wood has not been saturated with insecticide. The death watch beetle (*Xestobium rufovillosum*) is less common, but is still quite widespread. The larvae live for several years within the timber, and emerge usually in May to mate and lay their eggs. The death watch beetle lives commonly in dead wood such as occurs in the decaying crown of an old pollarded willow. It is said to fly in-frequently, but this is because it cannot take off unless the temperature is over 25°C, and in Britain it is seldom warm enough at the appro-priate time of year. Once I was laid up for a few days in a particularly hot May and was alarmed to find death watch beetles flying all round my bedroom – I had not realised that the house was so heavily infested. In hot years beetles flying in from outside no doubt start new infestations in suitable houses.

Old houses generally offer more opportunities to bats and birds to enter the roof space and enjoy its shelter. Old barns obviously provide more sites for owls and other birds than do modern buildings. In recent years architects have made considerable improvements in the appearance of farm buildings so that they are no longer the eyesores which were so often erected in the recent past but they have made little provision for wildlife. But even the most beautiful modern barn is no substitute for the tumbledown stone building erected in the fifteenth century. Fortunately, as far as birds, and bats, are concerned, nesting boxes may now be provided by farmers sympathetic to wildlife which does make it possible for some species to continue to live in the same area.

The old farmyard was generally infested with mice and rats. The ricks containing corn after harvest remained often well into the winter before it was threshed. Ricks often harboured hundreds of these rodents. They are not easily eradicated from new farm buildings, but they are more easily kept under control than in and around older structures. This population change is not one to which most conservationists will object. Many plants also find living room on buildings. Mosses, algae and lichens are found on old roofs and walls; the surfaces of many new building materials appear to be less easily colonised. Wallflowers and ferns are unlikely to be allowed to grow on new buildings.

Gardens in towns and suburbs support a surprising amount of wildlife. Eric Simms has studied the birdlife of Dollis Hill in the London area, and Denis Owen has recorded surprising numbers of insects in a garden in Leicester. Farm gardens, particularly when not kept too tidy, are likely to be even more productive, as there will be so much food available in the surrounding farmland. The old farm ponds were often so heavily polluted with the excrements of the livestock that they supported little wildlife, and it is possible that ornamental ponds created by newcomers living in houses no longer attached to farms may have greater value particularly for instance for frogs which may otherwise have no breeding sites on the farm itself.

It is clear that any oasis in intensively farmed arable areas will encourage the survival of many species of plants and animals. It has been found that churchyards can act as valuable small nature reserves.

However, it is not always easy to obtain the cooperation of the local village population. In one village I know well, the rector and parochial church council gladly agreed to allow a group of local naturalists to manage the churchyard. The churchyard was well supplied with trees, many covered with ivy, so it obviously had potentialities for wildlife. Nevertheless village people, and some who had moved away, objected to this 'desecration'. They were not themselves prepared to do any work to keep the churchyard tidy, and had it been completely left to be submerged with rampart briars and brambles they would probably have said nothing. But rough vegetation deliberately left for whatever purpose was obviously something which must not be tolerated, and so the project was abandoned. The only consolation is that funds were insufficient to have the whole area kept mown, and the objectors were too mean or too idle to do anything about it, so that the end result is not very different from that planned to encourage conservation.

CHAPTER 9

PESTS AND PESTICIDES

IF they were asked in what way modern farming was a danger to wildlife, most people would pick on the use of 'chemical sprays'. As the Royal Commission on Environmental Pollution said in 1979 in its *Seventh Report on Agriculture and Pollution*, 'The subject of pesticides is probably the most emotive of those that come within our study. The use of toxic chemicals by the farmer as the principal weapon against pest and disease attack and against weeds would commonly be regarded as the most worrying of the developments that characterise modern agriculture'. Rachel Carson's alarming book *Silent Spring*, published in 1962, alerted public opinion to the possible effects of pesticides on wildlife and on birds in particular. Its message is still widely remembered. Some naturalists believe that scientists are irresponsible in the ways they produce more and more deadly chemicals, that the chemical industry is criminal in the way it markets these substances, and that farmers are to blame for the way in which they slosh them about all over the countryside. Yet we have more song birds in many parts of Britain than we had twenty years ago, and pesticide damage to wildlife is, on the whole, decreasing rather than increasing year by year.

However, there is no doubt that pesticides can be dangerous to human health and to wildlife, and that their control is necessary. In a previous book in this New Naturalist series, *Pesticides and Pollution*, I described the situation in 1967. I now wish to show what has happened since that date, and to try to look at future prospects for the safe use of agricultural chemicals.

I shall use the term 'pesticide' for all chemicals or natural products used to control organisms which are harmful to agriculture. These will include insecticides used against insects, herbicides applied to control weeds, fungicides, acaricides, molluscicides, nematocides, rodenticides and other substances lethal to pests, to disease organisms, and to animals which transmit or harbour the organisms which cause diseases in crop plants or farm animals.

Pesticides may be harmful to wildlife in several different ways.

These chemicals are all, to a lesser or greater extent, poisons, and although many are particularly toxic to the target pests and comparatively less toxic to other creatures (thus some sprays against bean aphids are less lethal to bees which pollinate the bean flowers), nevertheless they are far from specific and so they cannot avoid damaging plants and animals within fields of crops receiving treatment. Secondly, pesticides may be applied carelessly, so that they damage wildlife in surrounding areas. Thus when a herbicide is applied, wrongly, in windy conditions, some is sure to drift on to roadside verges, hedges and nearby gardens. When pesticides are sprayed from the air, either by fixed-wing plane or helicopter, it is almost impossible to avoid contaminating some land outside the crop. Thirdly, chemicals properly and carefully used may be transported by living animals or other means to other areas where their effects may be serious. Thus a bird may eat grain treated with an insecticidal seed dressing in southern England, and then itself be eaten by a peregrine falcon as it flies across Scotland. The falcon may, as a result, be poisoned. Chemicals may be washed off the land into streams, and then be carried into lakes where fish may concentrate them in their bodies with fatal results. Some chemicals may be very toxic but unstable, when they will only have effects near the point of application. Others may be less acutely toxic but more chemically stable; these are the ones with long-term effects, sometimes far from where they were actually used on a farm. Fourthly, pesticides may alter the genetic makeup of pest and other populations, by killing only the more susceptible individuals, so that resistant strains are produced. This may present serious problems, so that pests become more difficult to control. We have little information about the effect of pesticide selection on non-pest species, though this also could produce important genetic effects.

There is some concern at the increasing number of pesticides available for use in Britain. In 1944 only 63 products were approved for use by farmers. The number had increased to 446 by 1956, and to 819 by 1976. However, there is little direct relation between the number of products available and their danger to wildlife. In fact it is true to say that some very toxic and persistent chemicals formerly used have now been withdrawn, and that many of those most recently introduced are of low toxicity. Clearly the effect of pesticides on wildlife will depend on the amounts of the various pesticides which are used, on the toxicity of these chemicals, and the ways in which they are dispersed into the environment.

Pesticides used in Britain are normally subjected to careful scientific scrutiny before they are available to farmers, horticulturalists and gardeners. We have the Agricultural Chemicals Approval Scheme, which is operated by members of the scientific staff of the Ministry of Agriculture, Fisheries and Food. They rely on the findings of the Advisory Committee on Pesticides, which has an important Scientific Sub-Committee, a body of experts including ecologists, toxicologists and other scientists, but excluding any representatives from industry. This Sub-Committee scrutinises the results submitted by any firm wishing to obtain approval for a new chemical. The firm must have made a thorough investigation in the laboratory and in the field, indicating the likely effects on wildlife as well as on crop pests. Eventually successful chemicals are included in the booklet *Approved Products for Farmers and Growers*, which should 'enable users to select, and advisers to recommend, efficient and appropriate crop protection chemicals, and to discourage the use of unsatisfactory products. The chemicals included in the Scheme are those used for the control of plant pests and diseases, for the destruction of weeds, for growth regulation and certain other crop protection purposes and for the control of insect and mite pests of farm stored grain.'

One interesting feature of the approval scheme is that it is voluntary, without any legal sanctions. Under it 'the manufacturers have undertaken not to market a product containing any new chemical for use in agriculture, horticulture or food storage, or to introduce a new use of a chemical already on the market, or to introduce a new formulation, until recommendations for safe use have been agreed with the Government Department concerned, if necessary on the advice of the Advisory Committee on Pesticides'. There are those who disapprove of this, and who wish strict legal controls to be introduced, as obtains in most other countries. I do not agree with this point of view. In some countries stringent regulations are not properly enforced. In others the strict letter of the law is not always obeyed, and those who wish to get round it can often do so. Our 'gentleman's agreement' seems to fit British conditions. Though not perfect, the situation regarding pesticides in Britain is better than in most countries. No new chemical which has caused serious harm to wildlife populations has been approved since the Pesticides Safety Precautions Scheme was introduced in 1957.

The scheme is quite flexible, and allows chemicals once approved to be given a second examination and if they have unexpected disadvantages they may be withdrawn. It also deals with the ways in which pesticides may be used, and with the stringent precautions relating to those which are very toxic, where the Health and Safety Regulations (which do have the force of law) apply.

I shall now discuss the ways in which the different pesticides which are widely used in Britain today may affect wildlife. I shall also try to forecast likely future developments in the use of chemicals in agriculture in Britain, and the ways in which our flora and fauna may possibly be affected.

HERBICIDES

Herbicides – weedkillers – are at present used by British farmers in far greater quantities than are all the other types of pesticides combined. In 1977 they used 17,000 tonnes of 'active ingredients' of all types of pesticides, and of this, 14,000 tonnes, more than 80 per cent, was herbicide. Of course the actual weight of material sprayed was much greater, for pesticide preparations when sold for use contain other substances, liquids or powders, which dilute the active chemicals and make it possible to spread small amounts evenly over a considerable area of ground. The formulation of the various products, marketed by different firms, is a highly skilled operation, and accounts for the fact that all preparations containing the same active ingredient are not equally effective, even if the differences suggested by the advertisements may be somewhat exaggerated. Many commercial preparations contain more than one pesticide, so one or more herbicides may be combined with a fungicide or an insecticide.

This enormous use of herbicides by our farmers is a recent development. Copper salts had long been known as toxic to weeds (as well as to fungal diseases, for which they were first used) but the amount applied to farmland was small. Sulphuric acid was used to burn off potato haulms to aid harvesting but for little else. The very poisonous substance sodium arsenite was used in gardens and glasshouses, and was the material most easily available to the criminal poisoner both in fact and fiction, but the total amount used, or misused, was small. Sodium chlorate was an efficient total weed killer for use on garden paths, though its liability to explode when dry was a serious disadvantage. Weeds in crops were generally controlled by cultivation, and by hand hoeing. Improved seed cleaning cut down the infestation of cereal fields by weeds like the corn cockle (*Agrostemma githago*), but a fair

range of wild flowers was generally tolerated and added to the beauty
of the countryside.

Although these weedkillers were toxic to wild plants and animals,
they probably had little effect, except very locally, on wildlife populations. Arsenic compounds did leave toxic residues in the soil, so
their action could be long-lasting, but there are few reports of any
serious effects. The total area treated was such a small percentage of
the whole that naturalists were quite unaware of any problem.

In the early 1930s dinitro compounds (e.g. DNOC) began to be
widely used to control weeds in cereal crops. These contact herbicides
acted selectively, killing broadleaved weeds with little damage to the
cereal (or to grass weeds) mainly because they ran off their surfaces.
DNOC is also used as a winter wash on fruit trees, when it acts as an
insecticide killing overwintering aphids, scale insects and red spider
mites. These dinitro compounds are very toxic to man as well as to
plants and insects, and can only be used by operators wearing protective clothing and taking other precautions. Any wild plants and
animals getting in the way of the spray are likely to be killed. Thus any
bird or mammal nesting or sheltering in a crop which was sprayed
almost certainly died. In the early days of the use of DNOC, spray
often drifted outside the target area and damaged both plants and
animals, but these local incidents did not appear to affect populations
significantly.

In 1942 an entirely different type of herbicide, the phenoxyacetic
acids, or 'hormone weed killers', started to be used and in a few years
largely replaced the poisonous dinitro compounds. They are effective
against most broadleaved weeds, but do not damage the cereal crops in
which they are growing. These herbicides soon became very widely
used, and continue to be sprayed in greater quantities than any other
group of chemicals. Weed grasses and wild oats are not controlled, and
at one time they presented serious problems in cereals. The most
widely used substance is MCPA. This, like most of the newer herbicides, is not very toxic to mammals and birds, being much less
poisonous than aspirin. MCPA did at one time do a good deal of
damage to wild flowers, as it was used on roadside verges to keep the
vegetation under control. Although we still have occasional incidents
when a stretch of verge is sprayed, most local authorities have agreed
to stop this practice. Spray drift onto wild vegetation is not uncommon, but the effects on wildlife on a national scale outside arable crops
would seem to be negligible.

As has already been indicated, these herbicides caused a substantial change in the weed flora of cereals, getting rid of charlock, poppies (*Papaver rhoeas*) and corn buttercups (*Ranunculus arvensis*) and replacing them with grass weeds, particularly wild oats (*Arvena fatua*). Some dicotyledons which are relatively tolerant of the herbicide also survived, including chickweed (*Stellaria media*), knotgrass (*Polygonum arviculare*), black bindweed (*Polygonum persicaria*) and fat-hen (*Chenopodium album*). Farmers found it necessary to introduce extra cultivations and break crops (which have their own weeds) to control weeds which flourished in spite of the frequent sprays of phenoxyacetic acids.

There was always the fear that the repeated use of MCPA might harm the flora and fauna of the soil, and so reduce its fertility. No serious effects on soil animals have been detected. With regard to bacteria, the most interesting development was their ability to 'learn' to break down the molecules of the herbicides. When a soil is first treated, the chemicals persist for several weeks, during which time they remain phytotoxic. After repeated applications the herbicides disappear more rapidly. This is because they are being increasingly metabolised by the soil bacteria. There may be other effects so far unrecognised on other micro-organisms and fungi, but in general the soil organisms do not seem to be seriously affected.

One chemical in this group, 2,4,5-T, deserves special mention. This is used as a brushwood killer, and is very effective against brambles and nettles. It is used in forestry, where it does not damage coniferous trees which are generally the main timber trees. At one time oaks were planted among conifers, with the intention of cropping the rapidly growing soft woods and then allowing the oaks to take over. There was then a change of policy, and it was decided to eliminate the oaks. This was done by spraying with 2,4,5-T; had the chemical not been available the elimination of the oaks would have been difficult and there would be many more growing to maturity in Britain today. 2,4,5-T also enjoys the notoriety of having been used by the Americans as a defoliant in the Vietnam war. This did serious ecological damage. Also the preparations used in Vietnam contained appreciable amounts of a very toxic impurity, dioxin, which can be lethal to man, and may also be teratogenic and carcinogenic. 2,4,5-T as marketed in Britain only contains a minute residue of dioxin, and although there is still public concern it is probably quite safe to use. It is one of the pesticides regularly used on nature reserves to control scrub growth. Clearly it could damage wild vegetation if carelessly applied, but otherwise it offers little danger to wildlife.

Where long-acting herbicides are required, for use on paths and railway tracks, substituted ureas and substituted triazines, particularly simazine, are available. Simazine is firmly held in the top 5 cm of soil. It is taken up by roots in this zone, and is effective for many months. It is also used to control weeds in orchards and asparagus beds, in maize crops and around broad beans. It is hardly at all poisonous to birds and mammals, and although it can make an area weed-proof for some months its effects on wildlife must be minimal.

The bipyridylium herbicides, paraquat and diquat, have been widely used in Britain since the late 1950s. They both destroy almost all green plant tissues by contact, with some slight translocation to other parts of the plants. This translocation is not generally very significant, and the results have been compared with those obtained by using a flame gun. These herbicides are rapidly adsorbed onto soil from which they are not released even when large doses are frequently applied over several years. Both chemicals are toxic to man – paraquat is one of the few agricultural chemicals which has killed people in Britain in recent years. Human deaths have occurred when children have, accidentally, drunk the concentrated preparation. Because of this danger paraquat is formulated in very dilute (and expensive) granules for use by the private farmer, the concentrate being available only to farmers.

Clearly paraquat will kill the above-ground parts of any wild plants which are sprayed. It is sometimes misused when it is applied to vegetation in ditches and hedge bottoms. The results are usually unfortunate. Grass is often eliminated, whereas perennial weeds with deep roots survive and take over the treated ground. I have already described (p.31) how paraquat is widely used prior to minimum cultivation of arable fields. This process caused considerable concern, but has been found to have little harmful effect on the soil fauna. In fact the earthworm population of fields treated with paraquat is much higher than that of fields normally ploughed.

I know of only one way in which paraquat, correctly used to destroy terrestrial weeds, may affect wildlife. It is usually quickly immobilised not only by soil but when it gets onto the surface of growing plants. These normally die in a day or two, and may then be eaten by herbivores without any ill effects. However, when sprayed onto wet plants the paraquat retains its toxicity, and there have been reports of hares (*Lepus capensis*) emerging from such fields with symptoms resembling myxomatosis. Their eyes and nostrils were swollen, and some of the animals died soon after. Other mammals and birds may have been similarly affected. However, such cases do not seem to be common, and the hare population as a whole does not seem to have suffered.

During the 1970s some 40 new herbicides were approved by the Advisory Committee on Pesticides and used by British agriculture. None of these is highly toxic to vertebrates; many are comparable as poisons to common salt and so unlikely to harm animal wildlife. Many are highly specific, able to kill weeds like black grass (*Alopecurus myosuroides*) and wild oats, which escape after the use of MCPA and other phenoxyacetic acid preparations. Recently barren brome (*Bromus sterilis*) has posed a new problem in some areas. This weed, which has spread from the hedgerows, is difficult to control, as many widely used herbicides are ineffective.

One of the newer chemicals which has given rise to some concern is asulam, which is used against dock (*Rumex* spp.) and bracken (*Pteridium aquilinum*). Used against dock in grassland it seems to be relatively specific, and many wild flowers, if they are present, survive quite successfully. It can be applied from the air to bracken in upland areas, being sprayed when the target plants are full grown. No symptoms are apparent during the rest of that year, but the growth does not resume the next spring and the underground rhizome rots away. Asulam, like most of the newer herbicides, has a low toxicity to vertebrates, other plants in sprayed areas generally survive, and side effects seem to be minimal. The main concern is that it is now practicable to eliminate bracken from huge rough areas where control was previously impracticable. Although most of us would welcome the reduction in the area of our uplands covered with bracken, we would not like to see this plant eradicated, and there is always the risk that cleared areas will be managed with little regard to wildlife. Asulam itself is not a danger, but it makes the profitable use of marginal areas a little too easy, and may thus encourage the exploitation of areas previously left largely to 'nature'.

Glyphosate, marketed under the name of 'Roundup', is another important new herbicide with remarkable properties. It is a broad-spectrum systemic herbicide, lethal to most higher plants. It is absorbed by the leaves and is freely translocated throughout the plant, moving down to the very ends of underground rhizomes. It remains active in the plants, but is inactivated almost immediately when it reaches the soil. Unlike paraquat, its toxicity to vertebrates is low. It has made the control of problem weeds like couch grass (*Agropyron repens*), bindweed (*Convolvulus arvensis*) and ground elder (*Aegopodium podagraria*) practicable. It could clearly be a danger to wild plants if used too freely, so perhaps the fact that it is still rather expensive may ensure that it is only applied to valuable farm land and private gardens where it can do little damage. It is unlikely to escape from its target site to affect vegetation elsewhere.

There are special problems when any type of pesticide is used in water, as the chemical may be transported by currents far from the target area. Herbicides, unlike some insecticides, do not become concentrated in living organisms, and are not transmitted in food chains, but their other effects can be devastating. Irrigation ditches are sprayed to allow the water to flow freely, and rivers and canals are kept clear to allow boats to pass easily without getting their propellers choked with weeds. The Ministry of Agriculture, Fisheries and Food issues official 'Guidelines for the Use of Herbicides on Weeds in or near Watercourses and Lakes', and the Advisory Committee on Pesticides only gives clearance to a limited number of chemicals for this purpose. Properly used these should not kill fish or aquatic invertebrates. The main danger is that they do their job so efficiently and kill reeds and floating vegetation. The dead plant material then decays and this may deoxygenate the water, producing results similar to those of gross organic pollution. This indirect effect of herbicide use can be of greater danger to wildlife than all the other effects of terrestrial weedkillers put together.

In general it would appear that any ecological damage done by herbicides is not because of their toxicity, but because of their efficiency. Farmers have always tried to grow clean, weedfree crops, and have generally succeeded by labour-intensive, cultural methods. Chemical herbicides have done the job more efficiently and with less effort. There is now the suggestion that it may not be an advantage to get rid of every single small weed. Some like low-growing pansies (*Viola arvensis*) and spurreys (*Spergula arvensis*) and annual grass (*Poa annua*) offer little competition to a vigorous crop and may even reduce erosion by heavy rain. These weeds also support a population of insects, which may be important food for young partridges (p.133) and other birds.

There is even growing evidence that some farmers are using more herbicides than are necessary, and in fact that after a few years of clean cropping, higher yields may be obtained if their use is temporarily suspended. The reason for this is that even the most selective weed killer does some – perhaps only slight – damage to the crop, and so it produces less grain. The difficulty is to know when to stop using these chemicals, and when to resume spraying, so most farmers prefer to keep up their application as an insurance policy.

Some organic farmers state that they can keep their fields reasonably free from weeds without using chemical herbicides. This is generally true, though I am doubtful whether the results are obtained because organic and not inorganic fertilisers are used, or because organic farmers generally eschew repeated plantings of the same crop on the same land, and use the type of rotation of different crops which was originally introduced largely to control weeds. There is little evidence that biological control of weeds in arable crops in Britain (i.e. the result of the activities of beneficial insects and other organisms) ever makes a significant contribution. There are many cases where weeds of rangelands in America and Australia have been controlled in this way, but only in large stretches of poor grazing where chemical controls are not economic, and where the beneficial insects have a perennial source of food so they can maintain and increase their population. This cannot happen in a cereal crop which may be sown in March and harvested only some five months later in August.

FUNGICIDES

Many crops suffer from serious diseases caused by parasitic fungi. The infamous potato blight which reduced the population of Ireland by more than 50 per cent in the middle of the nineteenth century was caused by the fungus *Phytophthora infestans*. Cereals suffer from seed- and soil-borne fungal diseases, from mildews and rusts, all of which can greatly reduce yields. Fruit trees may be subject to scab, mildew and canker, all caused by fungi. There is therefore a real need for efficient fungicides.

In the nineteenth century Bordeaux mixture, prepared by mixing copper sulphate, lime and water, was used to protect vines in Europe from fungal attack. It and other preparations based on copper were used on potatoes and other crops. Sulphur in various forms, as dusts and washes, was also used widely fifty years ago, as were some other preparations using mercury. Where copper was used year after year as in apple orchards, concentrations built up in the soil which lost most of its fauna and all of its earthworms. Other effects on wildlife appear to have been minimal.

Today the farmer has an enormous selection of fungicides at his disposal. These are widely used. Table 4 shows that in 1977 nearly a million hectares of cereals were treated with fungicides. Over 3 million ha were sown with treated seed; most of this was with fungicides. Of the other arable crops over 400,000 ha received seed treatment, and over 600,000 ha fungicides. Although these areas were so great, the amount

TABLE 4. Tonnes of active ingredients of various pesticides used on cereals and on other arable crops in England and Wales in 1977.

	Hectares treated	Tonnes of active ingredient
CEREALS		
Insecticides		
Organochlorine compounds	1,000	6
Organophosphorus compounds	294,000	107
Other insecticides	272,000	43
Seed treatments	3,358,000	48
Fungicides	978,000	588
Herbicides	4,408,000	8,026
Other pesticides	188,000	263
OTHER ARABLE CROPS		
Insecticides		
Organochlorine compounds	40,000	35
Organophosphorus compounds	274,000	99
Other insecticides	180,000	416
Seed treatments	430,000	2
Fungicides	616,000	882
Herbicides	918,000	6,131

NOTE: The areas treated refer to treatments. Some were sprayed more than once. Almost all the 3,600,000 ha of cereals were sown with dressed seed (3,358,000 ha), an operation which can only be done once! Some had no herbicide treatment, some had more than one spray, hence the total of 4,408,000.

of active chemical was relatively small, at least in comparison with the amount of herbicide used in the same year. For seed dressing the figure is exceedingly low. For other treatments the quantity of fungicide used per hectare is about a fifth that of herbicide for the same area. However, these comparisons have little real importance, for the toxicity of different agricultural chemicals differs so greatly. Fortunately many of the newest fungicides have a low toxicity.

A number of fungus diseases of cereals may be carried by the seed. The most obvious is bunt, caused by *Tilletia caries*, in which the grain is replaced by a revolting black ball of spores. This has been almost completely eliminated by using mercurial seed dressings. Various organomercury compounds are used. This has caused concern, because mercury, or rather some mercury compounds, are very poisonous. There have been many deaths in some underdeveloped countries where seed corn dressed with methylmercury dressings, supplied by relief agencies in time of famine, instead of being planted has been stolen and eaten by the hungry peasants. In Sweden birds have been affected when the same seed dressings have been used. The seed has been eaten by graminivorous birds, and the poison concentrated to dangerous levels when these have been consumed by predatory hawks. In Britain we have had no such trouble, probably because we use different chemicals, phenylmercury rather than methyl mercury compounds. Pheasants fed on treated grain in Britain did not seem to be affected. There are slightly elevated mercury levels in some wild birds, and scientists do not wish to see these increased. They are therefore glad that effective even less toxic alternatives such as carboxin and benomyl are now coming into use. However, the organomercurials are unlikely to have any serious long-term effects, as so little is used. Only about one milligram of mercury is added to every square metre of land when the seed dressings are used. Serious mercury poisoning of man and wildlife has generally been associated with the timber industry, where mercurial 'slimicides' may be discharged into rivers and lakes in much larger amounts. The danger of mercury is that in mud in such fresh waters bacterial action may transform relatively non-toxic chemicals in the discharges into very toxic methyl mercurials.

If they are not treated, cereal crops in Britain may suffer considerably from mildew (*Erysiphe graminis*), yellow rust (*Puccinia striiformis*), *Septoria* spp, and other fungal diseases. These can generally be successfully controlled by modern systemic fungicides. Progress has also been made in producing cultivars of cereals which are resistant to many of the fungal diseases. The trouble is that such resistance may not last for many years, as by mutation or selection the pest changes and overcomes it. Also strains of fungi which are not controlled by particular chemicals soon evolve. Some success in slowing down such evolution has been achieved by growing several different varieties of barley in the same field, one of the very few cases where any success has been achieved by trying to incorporate the ecological concept of 'diversity' into farming. But it looks as though research for newer and more effective fungicides will have to continue, to try always to keep one jump ahead of the pathogenic fungi.

Modern fungicides, being of such low toxicity to vertebrates and arthropods, and having so little phytotoxicity, probably have very little direct bearing on the flora and fauna. They make modern arable farming possible, and so contribute to its effects which have been described in earlier chapters. We have little knowledge of how wild fungi are affected. This may be more important than is at present realised.

INSECTICIDES

When people are concerned about the danger of pesticides to wildlife, they are generally thinking about insecticides. Yet in Britain, as Table 4 shows, we use comparatively small quantities of these chemicals. This country is lucky in that it is rare for any crop to be totally destroyed by an insect pest, something which commonly occurs in the tropics. If we had no insecticides farming could follow very much its present pattern. There would be quite serious losses of cereals, beans, vegetables and fruits, and the cosmetic quality of fruits and vegetables would be reduced, but our population would not starve.

Nevertheless the worry about poisonous insecticides has some basis. In the late 1950s many seed-eating birds were killed by eating corn treated with organochlorine seed dressings, and the population of the peregrine and other raptors which fed on these birds was considerably reduced. There have been numerous incidents when other insecticides have killed wildlife, and residues of toxic substances have often been found in wild birds at levels which may be dangerous.

Before 1939 several insecticides were in use in many parts of the world. Some of these were extremely poisonous. Thus hydrocyanic acid gas was used as a fumigant, and copper arsenite (in 'Paris green') killed caterpillars and beetles. Sodium fluoride was employed against ants which invaded houses, and nicotine to kill aphids in glasshouses. The most efficient, and safest, preparations were made of natural products, rotenone from *Derris elliptica*, and pyrethrum from the flowers of *Chrysanthemum cinerariafolium*. These vegetable products were comparatively non-poisonous to mammals and birds (though rotenone is used as a fish poison, its properties being to kill fish without rendering the catch dangerous to man). They break down to non-poisonous substances quite quickly, particularly if exposed to sunlight, and so are never long-term environmental pollutants. In fact the short life of pyrethrum was considered a serious disadvantage, and research was done to try to prolong its activity. The trouble about these vegetable insecticides was that they could only be produced in limited amounts, and that to increase the crop – and the yield – would have taken years to organise.

During the 1939–45 war two new groups of insecticide came into use. The first was the organochlorines, or chlorinated hydrocarbons, of which DDT is the most familiar. The second group, the organophosphorus compounds, was discovered in work with allied compounds like mustard gas which were being tested for use in chemical warfare. Parathion was the first member of this group to be widely used. The great virtue of these synthetic chemicals was that industry could, if need be, produce them in unlimited quantities.

DDT had been synthesised in the laboratory in 1874, but its insecticidal properties were not recognised until 1939. It was widely used during the latter years of the war, first to kill lice which would otherwise have caused a typhus epidemic in Italy, and then throughout the tropics against mosquitos which were vectors of malaria. I was involved in work with DDT during this period. It seemed then, and for several years after the war, to be the perfect chemical. It killed insect pests at dilutions which at that time seemed incredible, yet was practically harmless to man. In Naples the dust was liberally distributed among the underclothing of the population, many of whom did not change their garments for weeks at a time. No serious ill effects were reported. Volunteers went for days in underclothes impregnated with DDT, and while lice on their bodies soon died, none of the humans showed any untoward symptoms. Swamps were freed from mosquitoes using as little as a kilo of the chemical to the hectare. Pests of crops in all parts of the world were successfully controlled.

DDT is not entirely non-toxic to man. Its acute toxicity is about the same as aspirin. However, this is not a valid comparison, because small doses of aspirin taken daily are quickly eliminated from the body, while a substantial proportion of DDT is stored in the fat. Fortunately much of the DDT is transformed to its metabolite DDE, which is even less poisonous to man (though it may be more harmful to some birds, p.120). DDT has one interesting record, which is that, notwithstanding the large amount which has been used throughout the world, no human has ever died as a result of its use as an insecticide and millions of lives have been saved (although a death was reported when it was mistaken for flour and used to make pancakes). Workers in factories formulating DDT preparations have been found to have several hundred parts per million of DDT and DDE in their fat, without any noticeable ill effects.

It was not until the mid-1950s that there was any serious worry about the dangers of DDT and the other chlorinated hydrocarbons now in use to wildlife, though as early as 1945 the record of a meeting in London of the Royal Society of Tropical Medicine shows that I said that as it was such a powerful substance and so effective against pests we should be looking for its other ecological effects. In Britain in 1958 and over the next few years, large numbers of seed-eating birds were found dead in the fields in spring, and there were disturbing reports of a fall in the numbers of breeding peregrines and sparrowhawks, possible damage to golden eagles and buzzards, and numerous deaths of foxes and badgers. Eventually this was found to be caused by seed corn which had been dressed with aldrin or dieldrin as a protection against attack to the young seedlings by the larva of the wheat bulbfly (*Leptophylemyia coarctata*). In some years this was a serious pest, and dieldrin and aldrin had been found to be effective against it. What is ironical is that the damage was done as a result of an attempt to reduce the risk of widespread pollution. The insecticides were first broadcast over the whole area of the crop, using a kilo or more to the hectare. It was then decided to stick a minute amount of insecticide onto the grain, so as to concentrate it at just the spot where the pest attacked. This reduced the total amount of dieldrin or aldrin applied, but unfortunately it was in just the situation where it was ingested by seed-eating birds.

There was a comparatively happy ending to this story. On the recommendation of the scientific sub-committee of the Advisory Committee on Pesticides, it was agreed that seed dressed with aldrin and dieldrin would no longer be used in spring-sown corn. These dressings were allowed in autumn, when birds had other ample sources of food. This agreement was announced in 1962, which was incidentally before the publication of Rachel Carson's *Silent Spring*, and shows that scientists had been concerned with the danger of pesticides long before that book appeared – they had not, as it sometimes has been suggested, been blind to their dangers. The voluntary ban produced immediate results. The number of dead birds found in 1963 was small, and residues of the chlorinated hydrocarbons in wildlife started to fall. Unfortunately some farmers were prevented by bad weather from sowing their seed before the end of December and planted it in January or February, local bird kills did occur, and badgers which had consumed poisoned pigeons were also found dead or dying. Therefore the use of these chemicals as seed dressings has, by later agreement, totally ceased.

The greatest risk to wildlife from chlorinated hydrocarbons was found to occur in water. Fish and other aquatic animals are susceptible to all forms of water pollution, but particularly to pollution by fat-soluble poisons (like the chlorinated hydrocarbons) which may be concentrated from the water into the fishes' tissues by a factor of ten thousand or even more. The reason is that fish have to extract their oxygen from the water. Oxygen is not very soluble, so the fish must pass enormous amounts of water over their gills if they are not to be asphyxiated. While they do this, they also absorb other substances, particularly those which are soluble in fat. Fish-eating birds, such as the heron (*Ardea cinerea*), pick up this insecticide and may concentrate it even further.

A number of effects on wild birds were detected when the chlorinated hydrocarbons were widely used. Birds laid eggs with thin shells, which were more easily broken in the nest. Eagles and buzzards ceased to breed for several years, though they were insufficiently poisoned to be killed. Breeding seemed to be particularly susceptible to residues of DDE, which had previously been thought to be less toxic than the DDT from which it was metabolised in the birds' bodies. The population of the peregrine was reduced by some 75 per cent, the bird only surviving in the remote highland areas.

There was strong pressure from conservationists to phase out the chlorinated hydrocarbon insecticides as rapidly as possible. Farmers opposed this, as the chemicals were both cheap and effective. Farmers liked their long-lasting effects, conservationists feared that the chemicals might remain, and even accumulate, in the environment over periods of years. In the end in Britain it was agreed to restrict their use severely, completely withdrawing the most toxic substances endrin and heptachlor. The approved list still includes DDT, dieldrin and aldrin, each marked with a symbol indicating that the 'uses of this chemical have been limited by agreement under The Pesticides Safety Precautions Scheme'. Thus dieldrin may only be used as a dip against the cabbage root fly (*Erioischiae brassicae*) and DDT against some glasshouse pests. HCH, formerly called BHC, which is less persistent, is somewhat more widely used, and accounts for most of the 35 tonnes included in table 4, p.115.

The effect of the severe restrictions on chlorinated hydrocarbons has been generally satisfactory. Populations of peregrines have increased, though they are still well below the pre-war levels. Residues in sea birds have, for the most part, decreased. There has been some disappointment that residues have not fallen further, though levels seem well below those which damage the birds. The only disadvantage is that when non-toxic but persistent chemicals like DDT have been phased out, some of the replacements which are less persistent are more acutely toxic. In tropical countries human fatalities have occurred. Britain has been spared such tragedies, but there have been incidents when wildlife has been affected (p.122).

The first organophosphorus insecticides introduced to Britain after the war included parathion and TEPP. These are among the most poisonous substances in common use, and have killed thousands of operators throughout the world. In Britain strict regulations regarding protective clothing and the methods of application prevented more than a handful of serious accidents to man, but birds and mammals in sprayed crops and subject to drift in nearby areas were often killed. Fortunately these very toxic insecticides were not widely used in Britain, and were generally replaced by other OPs such as malathion which are several hundred times less poisonous. From the point of wildlife malathion, and also the very toxic parathion, have the advantage that they break down fairly rapidly and are not accumulated in animals' bodies.

Various systemic organophosphorus insecticides have been widely used. These include demeton methyl ('Metasystox'), sprayed in substantial amounts to protect beans and sugar beet from aphids. As a rule these systemic insecticides have little effect on wildlife. In beans ladybirds and other beneficial insects which help to control aphids are usually unaffected.

In recent years a considerable number of new organophosphorus insecticides and carbamates have been added to the approved list.

Carbamates, though chemically distinct, are similar in action to the OPs. These new substances have been used when pests have become resistant to older chemicals. Many of the newer insecticides are more persistent than the earlier OPs, though they are nothing like as long-lasting as DDT and the other chlorinated hydrocarbons. Some degree of persistence is necessary for seed dressings and other uses where repeated spraying is impracticable. Unfortunately some of the carbamates are highly toxic. Their use is at present restricted to only a few crops, but if they are more widely used wildlife is likely to suffer.

Seed dressings with aldrin and dieldrin have been replaced with some of the more persistent OPs including carbophenothion. I am surprised that this substance was approved for this purpose, as it is one of the more toxic of the new chemicals. I suspect that there was great pressure to obtain an effective replacement for dieldrin. In most situations the replacement has been safe and satisfactory, but it has unfortunately killed substantial numbers of greylag geese (*Anser anser*) and pinkfooted geese (*Anser fabalis brachyrhynchos*) in Scotland when they have grazed on cereal fields. Fortunately the incidents just described do not seem to have significantly affected the goose populations of Britain or of the world.

Natural pyrethrum has long been known to be an efficient insecticide, with the advantage that it is relatively non-poisonous to man and other higher vertebrates. Its disadvantage are its short life as an active compound, and the difficulty in growing the plants which produce it. For many years work on the synthesis of related compounds has been studied at Rothamsted Experimental Station, by M. Elliot and his colleagues. They have produced a number of promising compounds which are more active against pests than natural pyrethrum but which at the same time are even less dangerous to mammals. These will damage non-target insects, including beneficial species, but are likely to be harmless to other forms of wildlife.

Another line of research has been to try to produce chemicals which interrupt the development of pest insects. An insect produces in its normal life a juvenile hormone which delays the onset of metamorphosis. This, or a very similar compound, has been synthesised, and if applied at the right stage, when the insect is transforming from pupa to adult, the process is disorganised and the insect perishes. Great hopes were held out for this new approach to pest control. The hormone is not a poison, it does no harm to most other organisms, and is specific to the pest and its near relatives, so beneficial insects are not harmed.

Unfortunately the method has disadvantages. The chemical can only be effective for a very short period in the insect's life. The hope that resistance to the synthetic hormones would not develop have been confounded. It is therefore unlikely that this group of compounds will make any great contribution to pest control.

The Royal Commission on Environmental Pollution was greatly concerned with the development by pests of resistance to particular pesticides. Most pest populations contain individuals which differ in their susceptibility to an insecticide. There may be only a tiny proportion which is, naturally, relatively resistant. This may survive when the chemical is used. Nature abhors a vacuum, most pests breed rapidly, so a resistant population which cannot be controlled with this particular pesticide is produced. In practice this development of resistance is slow and by no means inevitable. Resistant insects, far from the popular conception of 'super pests' may be less vigorous than other members of the population, and unless all the remainder of the population is wiped out they will soon revert to being a small minority. But there are many pests in all parts of the world which are difficult to control, being resistant to several different chemicals. It is doubtful whether the chemical industry will always be able to keep up the process of inventing a new insecticide in time to control pests resistant to all those then in use.

In Britain resistance among agricultural pests is not yet a major problem. The Ministry of Agriculture, Fisheries and Food informed the Royal Commission that some ten pests in Britain are resistant to one or more insecticides; the most important is the peach-potato aphid (*Myzus persicae*), the vector of potato and sugar beet virus diseases. It is clear that resistance is most rapidly produced when pesticides are carelessly and excessively used. If farmers only use them sparingly so as not to try to eradicate the pests (when only resistant individuals will survive) but simply to reduce numbers below the level where serious economic damage occurs, then the insecticides are likely to remain useful for a much longer period. The subject is a difficult one, and we still have much to learn about the production of resistance, but if pesticide use is reduced then the damage to wildlife will be reduced also. The danger is that if a farmer finds that the normal dose of insecticide does not work he will then use much larger amounts with resultant pollution of the environment. This has happened in parts of Australia, where the boll weevil of cotton can only be kept in check with massive applications of DDT.

From the point of view of wildlife, it would be an advantage if insecticides could be done without, or used in much smaller amounts. Farmers are learning to revert to cultural techniques, different dates of planting, and other 'old' measures which were formerly their only means of reducing attack. As pesticides become more expensive, there is good reason to use them sparingly. Organic farmers believe that their methods reduce or even eliminate the need to use chemical pesticides – the few infestations that occur being curable with natural insecticides like pyrethrum and rotenone. I am not convinced that they have all the answers, but I have seen many cases where organically grown plants are less heavily infested than plants grown using chemical fertilisers in the same area. I have also seen some organically grown plants damaged by insects, so they are certainly not totally immune. I believe this is one of the problems which needs much more research – so far it has been ignored by most official research organisations.

There is increasing interest in biological control, where beneficial organisms play a major part in destroying pests. This has proved successful in glasshouses, where red spider mites are controlled by introduced predators. There has also been some success in forest and orchards, where trees support populations of beneficial insects which can build up over periods of years, and which can be in the right place when a pest arrives. I am doubtful whether biological control by predators or parasites will ever be of major importance in annual crops. Thus in a field of beans aphids may arrive and pullulate before a sufficient number of ladybirds can arrive to control them. In a garden, particularly if there are untidy clumps of nettles and other weeds for winter quarters, the ladybirds may be sufficient to control aphids on a single row of beans. If farmers could grow all their crops in tiny plots (something which would, economically, be quite impossible) they might be able to rely on the help of beneficial insects, but even then control might not be effective. The single row of broad beans in my garden is not always protected from aphid attack.

To sum up, I think that we do not need to be too depressed about the damage done to wildlife by insecticides. The situation in 1980 in Britain is much less serious than it was twenty years ago in 1960. We have an efficient system to control new insecticides, and although mistakes may be made these have been minor ones and possible to rectify. With a few exceptions the new pesticides of all types are less toxic and less persistent than those they replace. Where our wildlife is in danger it is from the general pattern of farming, in which pesticides play a part, but the chemicals themselves are not the main cause of the loss of wildlife.

Nematode worms form an important part of the soil fauna. As was noted in chapter 5, their total weight may amount to as much as 40 kg in a hectare of forest or grassland. Most of the nematodes feed on decaying vegetable matter, and play a part in the nutrient cycle in the soil. They thus contribute to soil fertility. Some species, however, are serious agricultural pests. They live as parasites on crop plants for at least a part of their lives. Some are even vectors of virus diseases. In the past the soil of glasshouses was commonly infested, with considerable damage to valuable plants. It is probable that in Britain eelworms cause a greater economic loss than do insect pests.

The potato root eelworm (*Heterodera rostochiensis*) causes losses to potato growers. Infestations become serious when potatoes are grown too frequently in the same fields. This means that farmers cannot always grow this valuable crop when they would wish to do so. A similar problem arises with sugar beet, where the parasite is *Heterodera schachtii*. Cereals and vegetables are similarly afflicted.

Many chemicals have been found with nematocidal properties, but they have in the past been too expensive to use on a field scale. Dazomet was used in glasshouses, but farmers could not afford the dose of some 350 kg per hectare which would have been necessary on a field scale. Recently other compounds, for instance oxamyl, have been found to be efficacious at much lower concentrations. Oxamyl sprayed onto the foliage is translocated downwards and affects the parasites in the roots. It is used at 1 kg or less per acre, which is a much more practicable proposition. It is unfortunately highly toxic to mammals, but is not cumulative so a series of small doses is not dangerous. If parasitic nematodes are killed, it is likely that free-living species will also be affected. Oxamyl is also insecticidal, so its widespread use would certainly affect soil fauna, and might damage many forms of wildlife. The effects of modern nematocides obviously need to be kept under scrutiny.

On farms and in gardens slugs can be serious pests. They flourish under damp conditions, and, unfortunately, they are numerous in soils where organic manures are used, as this provides an additional source of food. They attack potatoes and root crops, often reducing the value of the produce by their attack even when the total yield is not seriously affected. In autumn-sown cereals they may consume so much grain that resowing is necessary.

On a field scale slugs cannot be satisfactorily controlled although there are several chemicals which kill slugs in great numbers. These include metaldehyde, which has been used in gardens since 1936. Copper sulphate sprayed on warm evenings kills the animals which are on the surface at the time. The carbamate methiocarb is perhaps the most effective chemical, but its results are often disappointing. The majority of the slugs seem to survive, and rapid breeding replaces the losses. Slugs breed at temperatures only a few degrees above zero, and they are active, and feed voraciously, except in the coldest weather in winter. None of these treatments is completely selective, so wildlife could be affected as much as the pest species. I know of no reports of serious damage caused by these molluscicides.

Water snails (*Limnaea truncatulata*) are intermediate hosts of the liver fluke (*Fasciola hepatica*) which is a serious pest of sheep. Infection occurs in marshy land where the snails can flourish. Attempts have been made to rid such areas of snails by using copper sulphate, but these have seldom been successful. Fish, other animals and the vegetation suffer as much, or more, than the snails. From the conservation point of view the trouble is that the best way to get rid of the liver fluke, and at the same time to improve the grazing, is drainage. This has the effect of eliminating not only the snails but all the other wetland species. An effective and selective molluscicide might allow more marshes to be left undrained.

VERTEBRATE PESTS

Various mammals and birds damage crops and compete for food with farm animals. Their control is often difficult, and may be damaging to wildlife. The position is complicated by the fact that what one farmer accepts as harmless or even desirable wildlife another regards as dangerous vermin.

However, no one loves brown rats or house mice. These have been successfully kept down (though seldom exterminated) by the use of warfarin, an anti-coagulant which upsets the blood clotting mechanism so the animals bleed, internally, to death. As a rule the warfarin is presented in a bait which is laid so that birds and other animals will not take it. This is not always successful, and some other fatalities occur. The effects on wildlife populations do not seem to be significant. Unfortunately warfarin resistance has appeared in rats in several parts of Britain. Here acute poisons are used, with obvious risks to other animals. Attempts are now being made to control the grey squirrel with warfarin, something which was previously forbidden because of the risk to other animals. It is believed that if the warfarin-containing bait is only used in special containers which are only visited by squirrels this should be safe.

Woodpigeons damage brassica crops, and take a substantial part of the winter growth of clover leys. Attempts to control their number by providing cheap cartridges to farmers failed, though it provided sport for some country dwellers. Pigeon numbers were found to depend mainly on the food available, and shooting only killed off the surplus which would die of starvation anyway. Now attempts are made to cull pigeons using narcotic baits containing alphachloralase. Tick beans and other large seed are used as bait; these can only be swallowed by pigeons as they are too large for most small songbirds. In theory operators go round picking up the narcotised birds, and release non-pest species.

Foxes and badgers are animals to which there are somewhat varied reactions. In hunting country foxes may be killed by the hunt, but anyone who shoots or poisons them risks social ostracism. Badgers are a protected species, except where they are suspected of carrying tuberculosis (p.137). In this case they may be gassed with cyanide. Although this is performed by trained operatives, if a sett is also occupied by foxes they are likely to perish also. Deaths of other animals should be kept to a minimum.

Strychnine may be used to poison moles underground. The chemical is applied to earthworms which are inserted in the burrows. No other poison which is as efficient has been discovered. Unfortunately much of the strychnine supplied for moles is used, illegally, against other creatures. Every year foxes, badgers and even rare red kites are killed in this way.

Carrion crows are accepted as pests, and are given no protection under the various acts of parliament involving birds. They may be shot, and their nests destroyed. However, it is not legal to use poison baits against crows or other pest species. Nevertheless farmers commonly try to poison these pests, sometimes putting out eggs containing strychnine, organochlorine pesticides or other toxic substances. There have been instances where these poisoned eggs have been found and cooked by children. There have also been cases of poisoning of wildlife. Unless checked this might be one of the greatest risks to species like the red kite, where the total population is probably only about 30 pairs, and where any additional mortality could soon lead to extinction.

APPROVED NAMES OF CHEMICALS
MENTIONED IN THIS CHAPTER

Aldrin 1.2.3.4.10.10-hexachloro-1,4,4a, 5,8,8a-hexahydro-*exo*-1,4-*endo*-5,8-dimethanonaphthalene

Asulam methyl 4-aminobenzenesulphonylcarbamate

Benomyl methyl 1-(butylcarbomoyl)benzimidazol-2-ylcarbamate

Carbophenothion S-4-chlorophenylthiomethyl *oo*-diethyl phosphorodithioate

Carboxin 2,3-dihydro-6-methyl-5-phenylcarbamoyl-1,4-oxathiin

Dazomet tetrahydro-3,5-dimethyl-1,3,5-thiadiazine-2-thione

DDT (*pp'*-DDT predominant component) 1,1,1-trichloro-2,2-di (chlorophenyl)ethane (also known as dichlorodiphenyltrichloroethane)

Demethon-S-methyl ('Metasystox i') S-2-ethylthioethyl *oo*-dimethyl phosphorothioate

Dieldrin 1,2,3,4,10-hexachloro-6,7-epoxy-1,4,4a,5,6,7,8-8a,octahydro-*exo*-1,4-*endo*-5,8-dimethanonaphthalene

Dioxin (TCDD) 2,3,7,8-tetrachlorodibenzo-*p*-dioxin

Diquat 1,1'-ethylene-2,2'-dipyridylium bromide

Dinoseb 2-(1-methylpropyl)-4,6-dinotrophenol

DNOC 2-methyl-4,6-dinitrophenol

Glyphosate N-(phosphonomethyl)glycine

HCH (formerly BHC) 1,2,3,4,5,6-hexachlorocyclohexane

Malathion S-1,2-di-(ethoxycarbonyl)ethyl *oo*-dimethyl phosphorodithioate

MCPA 4-chloro-2-methylphenoxyacetic acid

Metaldehyde $C_8H_{16}O_4$

Methiocarb 3.5-dimethyl-4-methylthiophenyl methylcarbamate

Oxamyl NN-dimethyl-α-methylcarbamoyloxy-imino-α-(methylthio)acetamide

Paraquat 1,1'-dimethyl-4,4'-bipyridylium dichloride

Parathion *oo*-diethyl *o*-4-nitrophenyl phosphorothioate

Simazine 2-chloro-4,6-di-(ethylamino)-1,3,5-triazine

2,4,5-T 2,4,5-trichlorophenoxyacetic acid

CHAPTER 10

FIELD SPORTS

THE subject of field sports is one which arouses the emotions of countryman and town dweller alike in Britain. Some consider that field sports – or 'blood sports' – are an essential part of the tradition of the countryside. Others believe that they involve unacceptable cruelty and should be forbidden. Conservationists are equally divided. Some of the most active are also keen sportsmen, others consider hunting and shooting to be incompatible with the preservation of wildlife. I do not wish to enter into this controversy. My purpose is to consider how hunting, shooting and fishing are related to farming, and how these sports affect wildlife generally.

Most of the targets of field sports are wild mammals or birds, members of our native fauna, and so creatures which we wish to conserve. The populations in Britain may be entirely wild, as with foxes and hares, they may be introduced but feral, like many of our pheasants and red-legged partridges, the populations may be substantially augmented by birds hand-reared by man, as in pheasants, but they all rely on man to provide (or at least not to destroy) features in the environment on which they depend. From the point of view of conservation, by providing cover or food for animals and birds which are hunted, the numbers of many others which use the same facilities are maintained.

Some farmers are keen sportsmen, others are not. It is difficult to obtain accurate information on the proportion. In *Wildlife conservation in semi-natural habitats on farms. A survey of farmer attitudes and intentions in England and Wales* published in 1976 by the Ministry of Agriculture, Fisheries and Food, farmers in many parts of the country were interrogated regarding their attitudes to conservation. This report will be discussed in more detail in chapter 12. I am here concerned with the farmers' attitude to sport. Of 305 individuals who replied to the enquiry, 56 were considered to be 'sporting farmers' and 60 'naturalists'. It may be surprising to find that only 19 per cent of this sample were sufficiently interested in field sports to be prepared to make some small concessions in the management of their land to make

these sports successful – and that a slightly larger number was prepared to do the same for wildlife generally. These actual figures
probably do not represent accurately the situation throughout Britain.
The Game Conservancy reports suggest that it may be an under-
estimate, but there is no doubt that only a minority of farmers take any
active part in field sports.

Fox (*Vulpes vulpes*) hunting is perhaps the most glamorous of these
pastimes. There are 206 packs of foxhounds in Britain, with a total of
some 8,000 working hounds. In each year they kill nearly twice that
number of foxes – about 14,000. The supporters of hunting insist that
this cull is necessary, as otherwise the too numerous foxes would
become serious pests, and anyhow, they say, other methods of control
are even more inhumane. This is nonsense. In the parts of Britain
where hunting is most popular, foxes only continue to exist because of
the hunt. The site of every earth is known, and it would be possible to
gas them all with cyanide in a couple of weeks. I agree that gassing is
not a pleasant death, but it is quite quick. Once the fox was exterminated over a wide area, no further killing, by hunting, shooting or
gassing, would be necessary. At present fox hunting kills just enough
foxes to maintain a healthy population; it is a good conservation
practice.

Foxes are more difficult to control in other parts of the country. In
urban and suburban areas where they are now becoming common,
feeding on refuse and living in disused rail cuttings, derelict land and
old graveyards, hunting is impossible and shooting or poisoning is
difficult. In hilly regions the fox is more elusive, and is likely to survive
with or without man's help.

Foxhunting benefits other wildlife. Areas of rough ground, copses
and small woods containing earths are left undisturbed, so other
creatures also find a refuge. Hedges remain as jumps in hunting
territories, and barbed wire is discouraged as dangerous to horses. The
hedges are generally well laid and cut, so they are not as useful to birds
as overgrown ones, but they still do provide good nesting sites and
cover for insects and other animals. However, it is only the minority of
'sporting' farmers who will actively conserve foxes and so help other
animals. Those who are anti-blood spors do all they can to keep the
hunt away. They discourage foxes, destroy (as humanely as possible)
their earths, and probably support less wildlife than their sporting
neighbours.

The brown hare is involved in a number of sports. We have thirty

PLATE 17. Wild flowers are only found in undisturbed situations, not in arable crops or improved grass. *Above left*, verges of arable fields may support masses of primroses, which are disappearing from many parts of Britain; *right*, pasque flower, now becoming rare. *Below left*, spotted orchis; *right*, bluebells continue to grow for only a few years when old woods are felled.

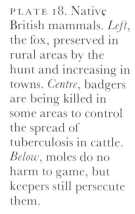

PLATE 18. Native British mammals. *Left*, the fox, preserved in rural areas by the hunt and increasing in towns. *Centre*, badgers are being killed in some areas to control the spread of tuberculosis in cattle. *Below*, moles do no harm to game, but keepers still persecute them.

PLATE 19. Introduced mammals. *Left*, the mink, an escape from fur farms, preys on native species. *Centre*, coypu, another refugee from the fur industry, is now an agricultural pest in East Anglia. *Below*, porcupine, recently escaped, damages young trees in Devon.

PLATE 20. *Above*, red deer are native to Britain and are now being domesticated for meat. *Below*, the roe is the other native deer; it flourishes in Forestry Commission plantations. There are probably more deer in Britain today than there were a hundred years ago.

PLATE 21. Amphibia and reptiles are at risk. *Right*, the common frog is now absent from much agricultural land. *Centre*, grass snakes are still found widely but in decreasing numbers. *Below*, the smooth snake is an endangered species and protected by Act of Parliament.

PLATE 22. Birds which have been affected by modern farming.
Top Left, sparrow-hawk, almost extinct in eastern England in 1960 due to pesticides; numbers are now increasing; *right*, barn owl, scarce in areas where nesting sites have disappeared. *Centre left*, lapwing, here nesting in field with corn shoots just appearing, flourishes with modern agriculture; *right*, skylark, here nesting in sugar beet field, also flourishes in modern large fields and is now a minor agricultural pest. *Bottom left*, red-legged partridge, here nesting in a pea crop, is doing well, unlike the grey partridge; *right*, pheasants have lost nesting sites with the removal of hedges, but numbers are maintained by rearing for the shoot.

PLATE 23. Birds which are increasing in numbers in Britain today.
Top left, osprey, extinct as breeding species before 1900, now nests in Scotland; *right*, bittern, returned to East Anglian marshes after some years' absence. *Centre left*, avocet, also extinct as breeder for many years, returned to Suffolk after the war; *right*, Savi's warbler, another successful recoloniser. *Bottom left*, black-tailed godwit, recently re-established breeding on Ouse washes; *right*, the great-crested grebe became rare in the nineteenth century but has increased greatly with increasing numbers of gravel pits.

PLATE 24. Positive steps towards conservation. *Above*, creating a mere at Holme Fen, partly to replace the lake drained for agriculture. *Below*, on another reserve where it has not been possible to use grazing animals the grass is being mown to prevent further encroachment by scrub.

packs of harriers, mostly followed by riders on horseback, sixteen packs of basset hounds and eighty-two packs of beagles, these two last being followed on foot. Hares are also coursed with greyhounds, and a great number is shot. There is a brisk export trade of hares to Germany, where they fetch many times the price realised in an English country market. The hare is something of a pest, eating growing corn and other farm crops. It is particularly damaging in young deciduous tree plantations, where it often bites off the leader from every one of a hundred or more young saplings at one meal, so that the trees grow up deformed. Hare numbers do not seem to be seriously affected by hunting or coursing, where quite small numbers are killed. Even hare shoots, which may kill many hundreds of animals in a day, only have very temporary effects. We do not fully understand what controls hare numbers. These increased sharply between 1953 and 1960, when myxomatosis (which did not kill hares) almost eliminated rabbits from many counties. It is generally considered that there was more food available for the increasing hare population. Since 1960 numbers have fluctuated, but have generally been near to, or below, the 1953 level.

Hares flourish in the open arable fields in East Anglia. They never burrow, and have their young in a nest (the form) which may be in grass or in a cereal crop. Some may be destroyed by cutting the grass or spraying with toxic insecticides, but the young are able to move out of the way soon after birth. Farmers interested in hunting or shooting hares do not need to make any special provisions for these animals, so other forms of wildlife do not gain, or lose, from their activities. Hare shooting like most sports is an inefficient method of pest control.

Deer are more numerous in Britain today than many people imagine for they are shy, elusive and mostly nocturnal. Only two species, the red deer and the roe (*Capreolus capreolus*) are truly indigenous. Fallow deer (*Dama dama*) which were introduced in mediaeval times are widespread, sika (*Cervus nippon*) introduced from the Far East, mix with the red deer in many hilly areas, while muntjac (*Muntiacus reevesi*) and Chinese water deer (*Hydropotes inermis*), the descendants of escapes from Woburn, occur in surprisingly large numbers over much of eastern England. There are probably more deer roaming about the British countryside today than have existed for several hundred years.

There are three packs of staghounds in the West country, hunting red deer on and around Exmoor. At present there is a fairly stable population of nearly a thousand deer, and the hunt tries to ensure that this is maintained. Present numbers are tolerated, particularly by the farmers who hunt; more would do unacceptable damage to the crops. Although by no means all the local farmers are supporters, the proportion seems to be much higher than that of 'sporting farmers' in most areas. When a stag is killed on farm land, the occupant is compensated for damage and generally receives his share of the venison. These deer could not live on farmland alone, they breed in woodland and on the moor, but without some winter feeding on the arable and grass fields they would find it hard to survive the winter. However much objectors may dislike staghunting, it has to be admitted that without it the red deer would probably cease to live on and around Exmoor. In the early nineteenth century when the pack was not in existence for some years, the population of deer fell to below fifty animals. The deer were killed by poachers and by farmers who, if they could not hunt, would not tolerate the damage done to their crops.

Fallow deer are hunted by the two packs of buckhounds in the New Forest. They kill few deer, the population being mainly controlled by shooting. Roe deer are also controlled by stalking. They are most numerous in Forestry Commission woodland, and if the numbers are too great the damage to trees is unacceptable.

Fallow and roe deer are often found on farmland, though like the red deer they also need woodland cover. Even red deer wander about the countryside of southern England in larger numbers than is generally realised. All deer, including muntjac and Chinese water deer, can damage crops, and may be shot if this damage can be demonstrated. Many farmers are quite happy to see a few deer on their land. Some encourage them, generally because they can then shoot some of the intruders.

The largest population of red deer is in the highlands of Scotland where numbers probably exceed 300,000. Here deer stalking is a lucrative commercial enterprise, and the venison, most of which is exported to Germany, is a valuable export. About 30,000 animals are culled annually. The deer compete with sheep for grazing, and may damage crops during the winter. The ordinary farmer is seldom involved in the deerstalking industry, and does what he can to keep the animals away from his land. As mentioned above (p.53) work on domesticating red deer is in progress, and they may replace sheep in some parts of Scotland in due course.

Pheasant and partridge shooting take place almost entirely on farmland in Britain. Some seven million pheasants are shot every year. Many tenants (who constitute about half the total number of farmers) do not own the sporting rights over their land, where they must allow the shoot to operate. They are generally cooperative, and are usually rewarded with a day or two of shooting towards the end of the season, and the present of the occasional brace of pheasants.

The modern farm, with few hedges, large fields and crops frequently subjected to toxic sprays, is not the most favourable environment for game birds. Pheasants usually nest in hedges and scrubby areas, so their numbers may be greatly decreased. Partridges nest in open fields, but early cutting of grass destroys their nests, and sprays reduce the amount of food available for their chicks. Where shooting is important, important modifications in the farming system may be made, in order to encourage the game birds.

Even in eastern England farms where shooting is encouraged tend to keep a larger proportion of their hedges. The value of the shooting rights may be thought to compensate, even monetarily, for the economic disadvantage of doing so. Even where most of the hedges are removed farmers plant up small areas with shrubs and trees for nesting sites. These are equally valuable to other forms of wildlife. They also plant small patches of kale or Jerusalem artichokes to provide additional cover. They cannot afford to grow uneconomic crops just to encourage game birds, but the needs of the shoot may persuade them to grow grass seed, which provides shelter and is not harvested until late July when the young birds are fledged, even if this is a little less profitable than another field of spring or winter barley. Sugar beet, oil seed rape and maize all provide favourable conditions during part of the season and so are favoured where game is preserved.

Pheasant numbers have kept up satisfactorily notwithstanding the changes in farming practice. This is partly because many pheasants are reared and released. The number of grey partridges has decreased greatly, and few are now being shot in many parts of eastern England where they were common less than fifty years ago. The reason for this is that the liberal use of herbicides has made the cereal crops much cleaner than ever before. At the same time few fields are undersown with clover and grass to produce a ley after the crop is harvested. This reduction in weeds and other plants means in turn that there are far fewer insects. These insects form the bulk of the diet of grey partridge chicks during their first weeks after hatching. This occurs at the beginning of June, and if the weather is cold and wet this, together with the shortage of insect food, produces a very high mortality. The red-legged partridge has survived better in most recent years, partly because it hatches later, and partly because the bulk of the chicks' diet from the start is vegetable. Sporting farmers are looking into the possibility of reducing their herbicide use, so that more of the less damaging weeds which best support insects survive. There is no doubt that the decrease in insect numbers in cereals must also have had a significant effect on other forms of wildlife, but so far population studies of other species have not been made.

Farmers and others trying to maintain large numbers of game birds have in the past been very concerned with the damage done by 'vermin'. This term is used in different ways by different people. The old-fashioned gamekeeper includes all types of predatory bird. So he shoots every hawk and falcon on sight. Even today some keepers continue this practice, ignoring the fact that the birds they shoot enjoy legal protection and that they kill more rodents and pigeons than pheasants and partridges. Every year we still have reports of deaths of rare raptors in illegal and cruel pole traps. As these birds breed slowly, and are present in small numbers, a few deaths could destroy the whole population. Fortunately most landowners and the more enlightened keepers no longer kill hawks. However, there is little doubt that control of some 'vermin' is essential if the maximum population of pheasants, partridges or grouse is to be preserved for the shoot.

Rats, which everyone accepts as vermin, are very damaging on lowland shoots. Where pheasants are reared rats may consume much of the food put out for the birds. It may be difficult to poison the rats without harming the birds, and traps may catch young poults. Rats breed so quickly that a few which escape control soon build up the population to pest proportions.

Probably the most damaging predator is the fox. It has been found to destroy as many as a third of the pheasant nests in some areas, and in most cases the hen bird is killed also. Rats destroy the eggs but seldom the hens. As many farmers who shoot also hunt, they are put in a very difficult position. Nevertheless keepers responsible for game birds shoot and snare a good many foxes every year, even if they try to keep this activity secret.

Crows, magpies, jackdaws and jays all eat game (and other) birds' eggs, and keepers try to control their numbers by shooting and nest destruction. Probably without this control these birds would be so numerous that they would damage other forms of wildlife. Gulls, which today are increasingly common in inland Britain, also kill some game birds and eat their eggs.

On a gamekeeper's gibbet, we often find desiccated corpses of stoats (*Mustela ermina*), weasels (*M. nivalis*), hedgehogs (*Erinaceus europeus*) and moles. Keepers believe that these all damage game birds. They probably eat some eggs. Stoats and weasels may kill young birds, but many scientists think that rats and mice are more often their prey, and that these predators may provide an overall benefit to the game. I am always sorry to see these wild mammals killed by gamekeepers. The stoat feeds very largely on rabbits, and its numbers fell alarmingly when myxomatosis eliminated this species from parts of the country. Hedgehogs are certainly not major predators of eggs. I cannot imagine that moles do these birds any harm. Fortunately many shoots cannot afford to employ full-time gamekeepers, and the work of vermin control is then largely restricted to discouraging crows and magpies, which are not endangered species, and which probably are the main culprits where game preservation is concerned.

Ducks and geese are shot by wildfowlers, many of whom are farmers. In fact in my experience genuine wildfowlers are the least bloodthirsty of men. Many shoot very few birds, and spend most of their time trying to produce better conditions where ducks and geese may feed and breed. They make more use of their binoculars than of their guns. Organisations like the Wildfowlers' Association of Great Britain and Ireland (WAGBI) on balance make a considerable contribution to conservation, though unfortunately there are irresponsible gunmen who go into the wetlands and shoot at anything that moves, who leave wounded birds to die, and who give a bad name to wildfowling. These ruffians are seldom good marksmen, so they miss their targets more often than they hit them, but they deposit substantial weights of lead shot in the marshes and this may poison other wildfowl which pick them up. Duck and swans dead from lead poisoning have frequently been found in eastern England.

Earlier I mentioned the way in which a few farmers have made ponds to encourage wildfowl, and how these may compensate for the loss of habitat caused by filling in ponds used for drinking by livestock and the drainage of wet areas on the farm. Some have gone to considerable lengths producing flighting ponds where duck may alight, providing nesting baskets for mallard which are virtually predator proof, and feeding the birds during hard weather. Quite large populations may be maintained; a 5 ha lake in the Netherlands contained over 400 occupied nesting baskets in the breeding season. In my own garden we provide no artificial nests for the wild mallard which frequent it in embarrassingly large numbers, encouraged by too liberal feeding by passing motorists when the ducks swim on the village pond, but in April we can usually find a dozen or more nests in shrubberies, herbaceous borders or compost heaps.

Geese may occur in such numbers that they overgraze grassland and cereal crops, but they usually do little damage and by encouraging the plants to send out more shoots they may contribute to higher yields in the next season. From time to time numbers are so high that cattle and sheep, in early spring, find little food left, but this is exceptional. Swans in East Anglia have also been blamed for overgrazing pasture in winter.

The most popular sport in Britain is fishing, with some three million anglers operating throughout the country. There is little relationship between farming and fishing, except that farm chemicals may cause pollution and kill fish. Anglers are often the first people to notice river pollution, and they make a considerable contribution to its control. They have had some influence on farming, by encouraging the safe use of pesticides and preventing these chemicals from getting into the rivers which they fish. Fish farming is now increasing in Britain, and this ensures the construction and maintenance of more ponds which support limited numbers of other forms of wildlife. The contribution to conservation is limited, as the fish have to be kept in rigorously controlled conditions, in as pure a culture as possible, to avoid diseases which may wipe out the breeding stock.

The general conclusion arising from this review of the main field sports in Britain is that they mostly contribute positively to wildlife conservation. Most of the species of mammals and birds which are killed by hunting and shooting nevertheless exist in larger numbers than would obtain were these sports to be eliminated. In most cases additional cover on the landscape of bare, intensive farms encourages other species of wildlife which would not survive in the bleak landscape.

CHAPTER 11

ZOONOSES AND DISEASES OF LIVESTOCK INVOLVING WILD MAMMALS AND BIRDS

IN this chapter I deal with problems arising from the involvement of wild mammals and birds in human diseases (zoonoses) and in diseases of livestock. Such diseases are, in general, more serious in the tropics, where plague from rats may still be a major human killer, and where trypanosomiasis from game may eliminate cattle over wide areas; but man and his animals may suffer appreciably in Britain and other temperate countries. As a result there are additional pressures on some members of our native fauna.

There has been much concern in recent years regarding the involvement of the badger in the spread of tuberculosis in cattle. Before 1939 tuberculosis, caused by the bacterium *Mycobacterium bovis*, was serious in British dairy herds, where as many as 40 per cent of the milking cows may have been affected. There was a small number of clean herds, where the farmer obtained a premium for his 'tuberculin tested' milk, but the spread of the disease to man (which had been common at earlier dates) was controlled by heat treatment of the milk ('Pasteurisation'). Nevertheless the disease was a serious one, some human infection still occurred, and it was harmful or even lethal to cattle, and therefore the attempts by the Ministry of Agriculture to eradicate it were generally applauded and (provided they received some compensation) supported by Britain's farmers.

Clearly the most important source of infection was from other cattle suffering from tuberculosis. In the mid-1930s it was estimated that over 20,000 animals had open lesions, and were discharging the germs into milk, urine, dung, into the air from coughing, or even from normal breathing. One animal in this 'open' condition rapidly infected a herd. The majority of infected beasts could not be shown to be spreading the bacteria in large numbers, but they still could be detected as infected reactors, a source of further potential danger. The policy was therefore to slaughter all these reactors. This seemed to be proving successful. By the 1950s 'open' cases fell to around 1,000, in the 1960s to single figures, and since 1970 they have virtually disappeared. The disease appeared to have been eradicated, or nearly eradicated, from most of Britain.

However, odd cases cropped up sporadically throughout the country. These were detected when the herds received routine tests, and the animals were slaughtered. The improvement continued, with virtual elimination of the disease, except in the south-west of England. In Cornwall, Devon and even as far east as Gloucestershire and Dorset, TB continued at an unacceptable level. The slaughter of cattle had failed to control it, as it had done successfully in other parts of Britain. There was clearly some other source of infection.

Farmers in Cornwall are said to have blamed the badger for some twenty or more years, but it was not until 1971 that scientists at the Ministry's Central Veterinary Laboratory at Weybridge isolated tuberculosis bacilli from a badger carcase from a farm in Gloucestershire. Since then many more badgers have been examined as well as other forms of wildlife from many parts of Britain. Badgers, particularly those from south-west England, have frequently been found to be infected. Infection has also been found in very small numbers of foxes, rats and moles, but these animals do not seem likely to infect cattle. The infected individuals were from farms where infected badgers were also present, and these were probably to blame. No infection has been found in deer, squirrels, hedgehogs, stoats, weasels, mice or voles.

Investigations showed that badgers are highly susceptible to bovine TB. It is probable that they were originally infected from cattle, and it is somewhat ironical that they are now being killed for passing back the infection. But there is little doubt that they are doing just that. An infected badger contaminates herbage eaten by cattle by urine – several million bacteria may be excreted at each urination – and by faeces, bronchial sputum and pus.

The danger appears to be greatest where the density of badgers is highest. In the south-west of England there are, on an average, about 40 badger setts per 100 sq km, but some areas have nearly ten times as many. These are, in general, the areas where TB in cattle has persisted. Although the badger is now protected by law, it may be killed by trained operators where there is reason to suspect it is spreading disease. Teams of workers have gassed many hundreds of animals during the late 1970s. There has been a substantial fall in the incidence of TB in cattle where badgers have been so controlled, though in some places, particularly Cornwall, the results are still rather disappointing.

There has also been a significant change in the pattern of the disease. In the 1930s, cows in milking herds were most commonly affected, as the disease was transmitted when these animals were in close contact at milking and when housed in winter. Today young stock, animals likely to graze on areas contaminated by badgers, most often show signs of infection.

The Ministry of Agriculture, Fisheries and Food, and most independent scientists who have studied the problem, accept as fact the part the badger has played in maintaining TB in some herds of cattle, and there is reluctant agreement to gassing setts in restricted areas. The official report states:

'The evidence accumulated during the period of this consolidated report (1976–1978) has established that the badger is the only species of wildlife which is a significant source of tuberculosis for cattle in Britain and that transmission of the disease from badgers to cattle has been shown to occur only in parts of the south-west region. The incidence of tuberculosis in cattle in the region has generally declined since action was taken against the badger. On the Dorset farm where investigations first began in 1974 the cattle have passed two successive tests since August 1977. In Gloucestershire and Avon there has been a marked reduction in the number of new herds in which reactors with visible lesions of disease have been found: in the Thornbury experimental area there have been no reactors with visible lesions of disease since February 1978. Only in Cornwall has this fall not been maintained but in spite of this the overall trend is encouraging and indicative of the progress made towards the Ministry's aim of bringing the level of tuberculosis in cattle in the south-west down to that obtaining in the rest of the country.

The incidence and severity of tuberculosis in the badgers examined at the Veterinary Investigation Centre, Gloucester has changed since the investigation began. There has been a reduction in the number found to be infected with bovine tubercle bacillus from approximately 20% on average for the 5 years prior to 1976 to 9.7% in 1978. There has also been a reduction in the number of carcases with lesions of disease and in the extent and severity of the lesions. The action taken by the Ministry must of necessity involve the killing of tuberculous and apparently non-tuberculous badgers in certain limited areas over varying periods: this should not adversely affect the future of the badger in those areas or elsewhere.'

This process of control is aimed, first, at eliminating all badgers which are infected with TB. This means that not only animals showing symptoms, but their immediate contacts which are likely to develop symptoms, a process which may take months or even years, have to be killed. Then the infected area must be kept free for a period, as bacteria remain active in cool damp situations for many months. After that it may be necessary to control the badger population to a lower level than obtained when it was a focus of infection, though there is not unanimous agreement on this point. But it is clear that as a whole it is only in the more crowded areas that badgers have been a danger.

It was only with reluctance, and after carefully weighing the available evidence, that conservationists, particularly those in the Nature Conservancy Council (the official body charged with safeguarding our wildlife), accepted the need for this controlled culling of badgers. There is natural revulsion at destroying a single individual of this attractive species, now the largest carnivore (though it is, from the point of view of its food, an omnivore) in Britain. It must also be admitted that some conservationists do not accept the 'official' view, and continue to oppose the Ministry's measures. However, the badger is not an endangered species, and it is likely to continue to flourish in most parts of Britain. There is probably a winter population (i.e. before the next generation of cubs is born) of something like 100,000 badgers in Britain, and the population is being maintained in most areas. If TB can be eliminated, this will be to the advantage of badgers as well as to cattle and humans.

Modern farming has reduced badger numbers anyhow, though rather less than might have been expected. In the mainly arable areas of East Anglia badger numbers are comparatively low, because of the shortage of sites for setts. However, the animals are opportunists, and refuse tips and old railway embankments often harbour flourishing groups. Unfortunately some farmers are prejudiced against badgers, and may seek any excuse to eliminate them from their land. They can do this without any hindrance if they 'improve' the area of a sett by removing banks, trees and shrubs and turning it into arable. Badgers are thought to be pests, they have been known to take chickens, to tread down or roll on cereal crops and to eat some of the farmer's produce. However, this damage is negligible, and insufficient to justify persecution, and we must not allow the danger of TB to be used as an argument to kill badgers except where the risk is a real one.

Another serious disease of domestic livestock, particularly cattle, is brucellosis, caused by bacteria of the genus *Brucella*. This is also spoken of as 'contagious abortion'. Man may be infected, often by drinking raw milk, and contracts undulent fever or Malta fever. The disease is being vigorously combated in Britain by slaughtering infected cattle and by vaccinating others, and it has been eliminated from many dairy herds. Although many species of wildlife including foxes have been found to carry the infective organism, they are thought to have picked this up from carrion, and there is no evidence of *Brucella* being given back to cattle – quite a different situation from that involving the badger and tuberculosis. At present, therefore, there is no move to kill wildlife as part of the brucellosis eradication scheme. However, there are different strains and species of *Brucella* with different infectivities to different species of mammal, and there could possibly be a situation where some wild animal might be responsible for reintroducing brucellosis into a clean area. Until this happens, there is no reason to include wildlife control in the attack on the disease.

Foot and mouth disease has caused serious losses to livestock farmers in Britain. It arises from infection by a virus similar to that of the common cold. Fortunately Britain is normally free from foot and mouth disease, and the Ministry of Agriculture makes every effort to perpetuate this situation. Outbreaks do occur, however, and in most cases the source is from imported animals suffering from infection, or from imported carcases, animal feeding stuffs, swill, packing materials and straw. To explain outbreaks of 'obscure origin', where there is no obvious source, it has been suggested that wildlife, particularly starlings migrating from infected areas in Europe, may be responsible.

Foot and mouth is a very contagious disease, rapidly spreading from one member of a herd to another. When a case is detected the area is strictly isolated, and all the cattle and other livestock, whether or not they show symptoms, are slaughtered and the bodies burned. It is often difficult to isolate an outbreak, and many surrounding farms over a considerable area may be affected. The huge bonfires cremating the corpses present a depressing spectacle, glowing into the night and emitting the unforgettable stench of charred flesh.

The disease is commonly thought of as mainly affecting cattle, but many other species are involved. In one period of twelve years, leading up to 1950, there were in Britain 223 outbreaks in pigs, 134 in cattle and eight in sheep. Wild ungulates are readily infected, and the virus has been found in birds, small mammals and even in hedgehogs. Man is not immune, though human cases, even in farmers dealing with sick animals, are not common.

My former colleague, the late R. K. Murton, made a particular study of the possible importance of starlings in transmitting foot and mouth disease. It had been shown that this was possible, for starlings ingesting the virus excrete it in an infective form in 10–26 hours, though the virus does not establish itself in the birds' tissues. However, his studies of many 'obscure' outbreaks in Britain caused him to come to the conclusion that birds were involved in very few, if any, of these. Particularly convincing is the finding that during the 1939–45 war, when imports of food and other normal vehicles of infection were restricted, but when there was a high incidence on the continent, and bird migration was unrestricted, obscure outbreaks in Britain were rare.

Deer are as susceptible as domesticated ungulates, and have been found to spread the disease in other countries. Attempts are made to control wild deer in Britain when outbreaks of foot and mouth disease occur, and deer may have helped to extend outbreaks in some instances. However, there is no evidence that deer serve as a reservoir, or that they have been responsible for primary outbreaks in Britain. There is reasonably good evidence that, on at least one occasion, a hedgehog was a focus of infection, and birds possibly play a part in spreading this disease once it is established. However, the general consensus of opinion would seem to be that wildlife is not important in the epidemiology of foot and mouth disease in Britain, and that there is no good reason for trying to eliminate any wild birds or mammals to prevent its occurrence.

Rabies in man and in many animals is a most unpleasant disease. In man it may not have a 100 per cent mortality, as is often stated, but if the nervous system is affected almost every patient dies a most unpleasant death. The treatment involves many painful injections, and sometimes allergies may arise so that the patient dies anyway. Symptoms may occasionally appear within 15 days of infection, but the incubation period is usually several weeks and may sometimes be as long as a year. Those bitten by rabid dogs are aware they are at risk, but infection may also result from handling an animal which may, at the time, show no serious symptoms. Wild animals with rabies may appear tame, thereby endangering children and others who handle them who thus do not realise that they are infected. In these circumstances the victim is unlikely to seek treatment until the disease has developed to an untreatable stage, when death is inevitable. Although only some 2,000 people die in most years, more than half of these in India, the high mortality and the horrific symptoms have caused man to be frightened of the disease and the authorities to be keen to prevent its spread.

Rabies was a serious disease in Britain in the past. It was probably introduced in the sixteenth century when it became endemic. Serious epidemics occurred shortly after its introduction and there were further major outbreaks in the eighteenth century and again from about 1870 until the disease was successfully eradicated in 1903. Although rabid animals have been imported since that date, the disease has, up till now, been kept out of Britain. Great efforts are being made to prevent any further importation of rabies; dogs and most other mammals must be kept in quarantine for many months and those trying to smuggle pets from Europe are, at last, severely punished. Nevertheless there is little confidence in official circles that rabies will not once again appear in Britain, and there are plans to deal with such a situation.

In the past rabies in Britain affected both dogs and other domestic pets, and also wild animals. All mammals and birds – all warm-blooded animals – can be infected, and can transmit the virus. When the main incidence of the disease is in dogs and other domestic animals, this is called 'urban rabies'. When wild animals are the main focus, this is called 'sylvatic rabies'. Before 1903 British rabies was mostly urban, though livestock and even deer in Richmond Park were infected; once this was understood, rigorous measures including muzzling dogs and killing infected individuals made eradication possible, though today we realise that there was an element of good luck in achieving this happy outcome. Sylvatic rabies is more difficult to deal with.

Since 1945 rabies has spread from Eastern Europe, where it has long been endemic, right across Germany and France to the coast of the English Channel. The disease has been essentially sylvatic, with wild foxes the main sufferers and carriers. There have been strenuous efforts to control the fox population on the continent, with some success. The disease has not been eliminated, its spread has continued, but human cases have, fortunately, been few.

If (many would say 'when') rabies reaches Britain, dogs will be strictly controlled, and compulsorily muzzled in the affected area. But the main target will then be the fox. Attempts will be made to exterminate foxes in a 'cordon sanitaire' around the focus of infection. This may or may not be successful. Fox control in rural areas of southern Britain may not be impossible, where countrymen know the location of every earth, but in mountainous regions foxes are more elusive. The urban fox, now common in many city regions, will be the most difficult to eliminate, and it will also be the most likely to transmit rabies. There is some hope that if foxes are reduced in numbers this may be sufficient to break the cycle and prevent them from transmitting the disease.

However, pessimists fear that these measures may fail, and that rabies may become endemic, when permanent controls and perhaps the use of compulsory vaccination will have to be enforced. Incidentally, Britain has so far not used vaccination as it is a second best method, and also it makes the recognition of the disease more difficult as vaccinated animals, when tested, may be confused with those developing an infection. Urban rabies has been controlled reasonably effectively in the U.S.A. using the vaccine, but sylvatic rabies persists in that country.

Rabies is most likely to reach Britain in a dog or other mammal, either smuggled by a criminally irresponsible visitor, or in an animal which is infected but which only develops symptoms after an unusually long incubation period – and this could be over a year. However, the virus could reach Britain without human assistance. Now that infection exists only 20 miles away across the Channel in France, it could be brought by migrating birds or bats. Birds are not often implicated in transmission, but over a period so many individuals fly back and forth that an unlucky accident is not impossible. Bats may be a greater danger. The bloodsucking vampire (*Desmodus rotundus*) in Central and South America is the main transmitter of rabies to cattle, and, on not infrequent occasions, to man. None of the 17 species of bat regularly found in Britain is a blood sucker, but all bats can easily be infected with rabies and may spread the disease accidentally if they are handled. In a serious outbreak of the disease there would be those who advocated trying to eliminate bats and other forms of wildlife which could possibly be implicated. Personally I would support fox control as necessary, dog and cat control as even more essential, and caution in handling all domestic animals as sensible, but I do not think that most wild mammals or birds deserve to be persecuted, at least as far as can be discovered from present knowledge.

Weils disease, or Leptospirosis, is possibly the most widely spread zoonosis in the world. The symptoms in man are headache, high fever, vomiting and, in the majority of cases, severe jaundice. The fever lasts for a week or ten days, and the patient may have relapses. Those most commonly affected are sewage workers, but infection on farms is not infrequent. In Britain the brown rat is the main carrier of *Leptospira icterohaemorrhagiae*, the organism causing the disease. This manifests itself when the germ penetrates man's skin or is ingested with contaminated food. Rat urine is the usual means of spreading the infection and anyone in a rat-infested environment is at risk.

As was mentioned in an earlier chapter, rats are not always accepted as 'genuine' British wildlife, as they were not present naturally when the country was separated from mainland Europe in the fifth millennium B.C. Rats were brought by man, unintentionally, and *R. norvegicus* probably did not arrive until the beginning of the eighteenth century. Large numbers of rats are infected and, where rats are common, Weil's disease is likely to occur. The obvious prevention is to get rid of rats, something to which few conservationists would object. But so far eradication has proved impossible except in isolated colonies. To get rid of the *Leptospira* would be even more unlikely, as an enormous range of mammals, birds and even reptiles have been found to serve as reservoirs of infection. But there is no evidence that *Leptospira* other than that from rats commonly affects man; so there is no reason to harry other wild creatures because of Weil's disease.

Many other diseases are associated with rats and mice. The most serious is plague. This, caused by the bacterium *Yersinia (Pasteurella) pestis*, is essentially a disease of rats and other rodents, carried from rat to rat by fleas. Man is only infected when he accidentally picks up an infected rat flea. Although plague is not endemic in Britain today, it was the cause of the Black Death in 1348, when nearly half our population died, and of the Great Plague of 1665 which killed 60,000 of London's 450,000 citizens. Plague occurs most readily where man is in contact with the black rat, a species whose fleas are efficient transmitters of *Yersinia*. The black rat is again not truly British, it was probably brought here, and its fleas with it, in the twelfth century. (Some zoologists think it arrived much earlier, but there are no records of epidemic plague much before the Black Death and I am inclined to accept the twelfth century arrival date.) The black rat is seldom found in rural areas today, and rarely infests farm buildings. There is always a risk of the brown rat or of other rodents spreading plague again, if the disease entered Britain (which, in these days of rapid travel is always possible), so the reasons for trying to eradicate rats are reinforced. But once again other wildlife is unlikely to be involved to any serious extent.

There are many other diseases of farm animals and even of man in which wildlife could play a part. Avian tuberculosis has been found in as many as 4 per cent of wood pigeons, and infection in man has been reported. However, this is an uncommon condition (in man) and is not thought to warrant more strenuous control of pigeons than is already, rather inefficiently, in force. Salmonellae (the organisms commonly causing 'food poisoning') have been isolated from most living animals at some time. Both invertebrates and vertebrates have been involved. In most cases there has been little danger to man, but there are reports of hares and pigeons feeding on salad crops and contaminating these with salmonella in their excreta.

Psittacosis or 'parrot disease' causes a severe and sometimes fatal pneumonia in man. It (or more generally 'ornithosis') is caused by the microorganism *Chlamydia* which is widespread in wild and tame birds, most of which show no symptoms. Parrots and budgerigars often infected their owners until importation was more carefully controlled. The pigeons in Trafalgar Square are infected, so tourists who handle them are at some risk. Many wild birds, particularly sea birds, carry the infection. Human cases have been reported on farms, where chickens are involved, and from the Faroe Islands, where fulmar petrels were implicated. Fortunately when derived from wild birds the disease in man seems usually to be less serious than when parrots are the source.

Farm animals suffer from diseases caused by a variety of helminths, including liver flukes and various tapeworms. The liver fluke, *Fasciola hepatica*, is a serious pest of sheep. It only occurs in wet pastures where the intermediate host of the parasite, a water snail, is present. As a rule sheep are the main cause of infection to snails, and control is by attacking the snail population either by using molluscicides or by draining the pasture and so eliminating the snail (and much other wildlife besides). But many wild animals including deer are also host to liver flukes, and could act as reservoirs if the disease were wiped out in sheep. At present these wild reservoirs are not thought to be important.

Tapeworms not only affect farm animals, but may reach man when he eats meat that has not been sterilised by thorough cooking. Here domestic animals, particularly dogs, are most likely to bring infestation into the home. Once again wild ruminants may serve as a reservoir, though farm animals are most likely to be the source of infection.

Troublesome arthropods which attack man may be normally associated with wild mammals or birds. In Britain these do not usually carry disease, as they do in tropical countries, so are generally not considered important, though they may cause serious discomfort. Thus the harvest mite *Trombicula autumnalis* is normally dependent on the rabbit in Britain, though it can also live on small rodents. Troublesome infestations are commonest on chalky soils where rabbits are numerous. The adult of this mite is free-living, and vegetarian. It lays its eggs in the soil. The larvae attach to the skin of a mammal (the ears in the rabbit), where they remain for some days (unless scratched off) engorging themselves on its serum. In man they cause severe irritation, so much so that they are usually scratched off and die, leaving a weal which may be slow to heal. Fortunately the British mite does not carry any disease, unlike the related mite which carries scrub typhus in S.E. Asia.

This catalogue of possible associations of wildlife with animal and human diseases is far from complete, but the major dangers have been included. My general conclusion is that at present the badger is important, in a few very restricted areas, for spreading bovine tuberculosis. Were rabies to reach Britain, fox control would be essential. Otherwise wildlife is of little importance, though further research is needed before we can be quite sure that every other species has a clean bill of health. Also we must be vigilant, for new parasites might prove more dangerous and incriminate further species if these were found to act as hosts.

CHAPTER 12

WILDLIFE CONSERVATION ON THE FARM

I T is clear that modern farming is generally harmful to wildlife. On the other hand, as over 80 per cent of the surface of Britain is farmed in one way or another, unless our native flora and fauna can survive on farmland many species will die out altogether, and most of our people will never have the opportunity of seeing wild plants and animals. Only by obtaining the cooperation of farmers will this survival be possible. How then can we persuade the ordinary farmer that conservation is worth while?

The Nature Conservancy Council, in its excellent booklet *Nature conservation and agriculture*, published in 1977, sets out some of the arguments, as follows:

'Why does wildlife matter? All living organisms are interrelated; all crops and domestic animals are descended from wildlife, and all depend on wild species directly or indirectly. In particular, most organisms depend upon the ability of green plants to use energy from the sun, and most of the plant material of the planet is made up of wild species. All green plants and all animals rely on micro-organisms for cycling the chemical elements necessary for life, while many plants are dependent upon insects and other animals for fertilisation. In some ways the totality of living organisms resembles a living body: each part is dependent upon other parts. However, as in a body, some of the parts are more important than others. There are probably at least two million species of organism in the world today; many of these cannot be essential for the continuation of life on the planet, and some are harmful to man. On the other hand, we do not know enough to say for certain which species are essential, nor do we know which ones would be most valuable in providing material for evolutionary processes in the future. The Nature Conservancy Council believes that the greater the diversity of genetic material the greater are the opportunities for evolution and the deliberate breeding of plants and animals by man. Generally speaking, therefore, it is wise to ensure the survival of as many species as possible. This is what conservation is about – maintaining biological diversity and so keeping the options open as wide as possible for our own and future generations. The value of maintaining wild stocks of domesticated species has been demonstrated by plant breeders on many occasions in the past. The need for genetic material for such purposes and for domesticating other plant and animal species is likely to grow as fossil fuels become exhausted and fertilisers and pesticides based on them become increasingly expensive. Quite apart from this, modern man still depends on green plants and bacteria for his survival, however much technological societies may forget it. While man depends upon the survival of some forms of wildlife, he also requires the rigorous control of others, notably the vectors of disease and competitors for food and other products of use to him. Further, the control of the numbers of some abundant species is often necessary for the sake of other more vulnerable wildlife species. Indeed such control is an integral part of optimising biological systems and hence of conservation in its widest sense.

'Man does not live by bread alone. From the earliest times he has struggled against nature, but he has also derived deep-rooted satisfaction from his kinship with it. This has been expressed in art for thousands of years and more recently in literature and scientific research. Today millions derive pleasure from their contact with nature both in their work and on holiday: people in modern societies wish to have some contact with wildlife and their need seems to grow with increasing industrialisation and prosperity. Many of the rare, spectacular forms of wildlife are valuable for cultural reasons. Wildlife is not an optional extra. Losses of the rarer species would greatly impoverish the cultural life of man, and losses of many abundant ones would put our survival in jeopardy. Wildlife does matter, and so measures must be taken to conserve it.

'Conservation and agriculture are interdependent. Agriculture depends upon the conservation of beneficial bacteria, soil invertebrates, pollinators, predators and parasites, while the conservation of most other species depends upon what agriculture does or does not do to rural land. Further, some wild species will almost certainly prove to be valuable to agriculture in the years ahead.

'The conservation of the species currently necessary for farming is assured to some extent by a built-in safeguard: if a farming practice is detrimental to productivity the farmer is likely to notice and to take measures to correct it. On the other hand, the effects of farming practices on wildlife outside the cropped areas will not affect the farmer's livelihood and so may go undetected. For this reason most emphasis in this paper has to be put on the conservation of the flora and fauna on the uncropped areas of the farm. This should not be allowed to obscure the obvious fact that the conservation of the species necessary for farming is essential and provides much of the common ground between agriculture and nature conservation.'

Everything in the statement is true, but, in the context of farming in Britain, I find parts of it unconvincing. On a world scale what is said about the possible value of wild plants and animals in future agricultural developments is unexceptionable. There are many wild species of plants which may one day be the parents of new crops, or may contribute to the genetic make-up of cultivars with resistance of pests and disease. There are relatives of the potato in the High Andes in this class, and grasses in many parts of the world which may contribute to new strains of cereals. There are wild animals which may be domesticated, and produce more efficient conversion of vegetable foods to meat than most existing livestock. I fully agree that we must make special efforts to preserve really large areas of the remaining natural ecosystems, such as the tropical forests of South America, first because we know so little about the plants and animals of these regions, and secondly because their destruction may affect the climate of large parts of the globe. But I cannot in all honesty say that I believe that these arguments apply to the areas and the species I wish to safeguard in Britain.

Even in other parts of the world we must be careful not to overstate the conservation case. I should deeply regret the extinction of any of the large mammals, rhinoceros, giraffe, lion, tiger which still exist in reduced numbers in some tropical countries. I think that it is right that we should try to ensure their conservation. But as the human population grows, and more food production is necessary, none of these animals is likely to survive in the truly wild state. They do nothing but harm to the local farmer, who cannot permit them to continue to rove around his land, destroying his crops and killing his livestock. The only hope for these large species is that they will survive in zoos or in special parks where they can continue to live in what appears to the tourists (who pay for their upkeep) a natural manner, but one which is only sustainable if the population is managed by park rangers and the surplus population is scientifically culled.

I am glad that the Nature Conservancy Council is careful not to appear to support the view expressed by many sincere conservationists, and by some who call themselves ecologists, that if only farmers took steps to ensure the conservation of wildlife on their farms, they would be more prosperous, obtaining better yields and suffering less damage from pests and diseases. If only this were true, there would be no opposition to conservation! Unfortunately it is not. Good farming, with the maintenance and enhancement of the fertility of the soil, so that sustainable and even improved yields of arable crops are obtained, is possible when wildlife is reduced to a minimum. Livestock will generally do best when there is no competition from wild species, when they have the sole use of the food supply and do not pick up diseases from wild sources. In almost all cases conservation will cost the farmer money, and will seldom increase his income.

Why then should farmers encourage, or even tolerate, wildlife on their land? The Nature Conservancy Council statement gives one answer. It points out that 'man does not live by bread alone'. He appreciates wild plants and animals for artistic, aesthetic and even spiritual reasons, as objects for scientific study and for the pleasure and education of his children. Without wildlife Britain would be an impoverished country. If we are really honest, we have to admit that we wish to conserve wildlife because we like it. Can we persuade farmers to like it too?

The thing which surprises me is that farmers far from being a lot of materialistic philistines, eager to destroy every unprofitable plant and animal, are, on the contrary, very often highly sympathetic to conservation. We saw from one survey that a fifth of one sample of farmers, a greater number than those engaged in field sports, had sufficient interest in natural history to be prepared to make some modifications of their farming plans to encourage wildlife. In the government white paper, *Farming and the Nation* presented to Parliament in February 1979, we read:

'The Government recognise that in certain special cases the requirements of agricultural production must take second place. For example, the main recommendations of Lord Porchester's Study on Exmoor have been accepted by the Government, including the objective of securing in the national interest the conservation and management of defined areas of moor and heath of exceptional value within the Exmoor National Park. More generally, in considering any application for a grant for capital equipment or land improvement the Agricultural Ministers in Great Britain are bound by the Countryside Acts to have regard to the desirability of conserving the natural beauty and amenity of the countryside. As a result a number of schemes have been modified and some grant applications have been refused. Similarly, while land drainage is recognised as an important means of increasing the productivity of agricultural land and providing protection from flooding, water authorities and Ministers are obliged under the Water Act 1973 to take into account any effect which drainage proposals would have on the environment and conservation.'

Thus it is official government policy to give precedence to wildlife in some areas where a convincing case can be made.

There have been many exercises to study the problems of farming and wildlife. The first important breakthrough was the famous 'Silsoe

Conference' held on the 9 to 11 July 1969. This arose from the initiative of the Royal Society for the Protection of Birds, which made an approach to the government's National Agricultural Advisory Service (NAAS) in 1967. As a result a formidable group of ornithologists, conservationists, farmers and agricultural advisers came together for a weekend when they discussed the subject with particular reference to one area of 150 ha of farmland in Hertfordshire. The published account of their deliberations *Farming and Wildlife: A Study in Compromise* sets out for the first time the ways in which the differences between farmer and conservationist may, to some extent, be reconciled. The delegates accepted the need for farming to be financially profitable, but assumed that farmers could be asked to forego some of the advantages of the rigorous application of modern intensive methods. The conservationists agreed that the number of hedges must be reduced, but tried to identify, and preserve, those of greatest value to wildlife, and they also suggested additional planting of copses and shelter belts which would not seriously endanger the efficiency of the farm while at the same time providing useful cover. A reasonable policy for pesticide use was also worked out, controlling major pests with the minimum damage to non-target species. The Silsoe conference undoubtedly demonstrated the possibility of ensuring that at least some wildlife could continue to exist under modern farming conditions.

Some people feared that the Silsoe exercise would simply be a useful public relations exercise, with no long-term effect. Fortunately it proved only to be a beginning. Thus the Minister of Agriculture, Fisheries and Food proved to be unexpectedly sympathetic to conservation, and said in January 1970: 'I intend to introduce more emphasis on the opportunities for conservation in the training of my advisory services. New courses designed to achieve a broader understanding of conservation management and its relationship with farming practices are already being planned. As a result, advisers will be more easily able to suggest to farmers and others in what way and at what cost conservation interests can be safeguarded. This is a practical contribution which I intend to make to European Conservation Year' (1970 had been so designated).

Another important development was the Farming and Wildlife Advisory Group, with members from all the major farming, landowning and conservation bodies. There is a central Group, based on the headquarters of the Royal Society for the Protection of Birds, and no less than 32 local county groups. There are representatives from official bodies, the Ministry of Agriculture, the Forestry Commission, the Countryside Commission and the Nature Conservancy Council. The Country Landowners Association and the National Farmers' Union send members, as does the Royal Institution of Chartered Surveyors. The conservation side includes the Royal Society for the Protection of Birds, the British Trust for Ornithology and the Society for the Promotion of Nature Conservation. The groups are all completely independent, and are not controlled by the parent bodies. The groups have done a great deal to help farmers who wish to make some positive contribution to conservation on their land.

There have also been a great many more 'Silsoe type' exercises. Accounts of eleven such have been published, and details are given in the bibliography. These covered much of Britain, areas involved being Dorset, Berkshire, the urban fringe, Northumberland, Essex, Cheshire, Wiltshire, Suffolk, upland Wales, Leicestershire and, again in Wiltshire, on the farm of the Royal College of Agriculture, Cirencester. A great many other local conferences have not given rise to formal publications.

Various competitions have also taken place. The journal *Country Life* offered its award to 'the farmer in the U.K. who, in the opinion of the judges, has done most to encourage wildlife conservation on his farm within the constraints of successful commercial farming'. The first prize went to a farm of 30 ha operated by a tenant who kept no fewer than 68 cows on this limited area. One of his most successful schemes was to plant a former rubbish dump with a selection of native trees, and to incorporate a pool fringed with willow and alder. He was particularly praised for the way he showed that really intensive farming need not entirely eliminate wildlife. The two joint second prize winners operated much larger farms in Wiltshire.

In the early spring of 1980 I had personal experience of a slightly different type of exercise. The East of England Agricultural Society organised a Young Farmers' Club Farm Management Competition. Some sixteen teams of 'young farmers', six young men and women per team, spent a Sunday on a large farm on the borders of Northamptonshire and Leicestershire. They had to prepare three reports. First, they had to value a selection of livestock and farm machinery. Secondly, they were given the task of preparing a management plan for the farm, giving reasons for modifying it from that now in operation. And thirdly they had to suggest ways in which wildlife conservation could be developed without damaging the overall profitability of the enterprise. Judges with appropriate experience were appointed for the three sections; I had to judge the conservation report. We started with a short speech from the Hon. Richard Butler, President of the National Farmers' Union, and I was gratified to hear him stress the importance of conservation, and of the way in which farmers should think of themselves as trustees of the land, of the beauty of the countryside and of the wildlife living on it.

We judges were given a detailed tour of the farm while the teams made their own examinations and prepared their reports. These we received during the afternoon. I was most impressed with the results. One team, from Ashby-de-la-Zouch, put in a model report. They had identified the various areas of special wildlife interest, where normal farming (in this case a mixture of arable crops and livestock) was difficult. These included a mediaeval quarry (a deep hole with water at the bottom), the banks of a river and a tributary stream flowing through the farm, some small fields of very old grassland, and a disused railway line much of which had been incorporated into the field system, but where a deep cutting and a high stretch of embankment remained. For all these areas sensible planting schemes with native trees and shrubs were suggested. The farm still retained the main system of well-maintained hedges around its large fields (the farmer was a keen fox hunter) which the young farmers said should be retained and laid when required. They suggested some further minor planting of cover and shelter belts where these would not prejudice efficient farming, and they made the appropriate proposals about the careful and conservative use of pesticides. As far as conservation was concerned, the Y.F.C. of Ashby-de-la-Zouch had done their own Silsoe exercise, and done it as efficiently as the experts could have wished.

The other reports were equally interesting. Five teams obviously knew what they were about. Though the reports were not as comprehensive as the one which I have described, they covered most of the same features. Some missed the mediaeval quarry which was at the top of a hill rather a long walk from the road. Some did not make sufficient use of the river and its banks. But they all made useful, practical suggestions. The other half of the reports were of a very different calibre. Most of them were brief, they contained a few generalisations about the value of wildlife, some identified features for planting, and there were the expected comments about pesticide use, but the writers had obviously taken little real interest in conservation.

I found the results very encouraging. Nearly half the teams clearly understood what the exercise was all about. The fact that the rest did not should not surprise us – if the others can be so well educated, so in the future can they. Incidentally, I was further encouraged when I found that the farmer himself had already included many of the main recommendations, such as tree planting in the railway cutting, in his own future plans.

Nevertheless, we must not be *too* optimistic. It is still only a minority of farmers that cares about wildlife and its conservation. The others do not come to our meetings or take any part in 'Silsoe-type' exercises, they do not even know that there are voluntary bodies such as the County Naturalists' Trusts or official organisations like the Nature Conservancy Council. Some are converted when their children learn about wildlife and persuade their parents to take them to visit nature reserves when these are open to the public and when simple but interesting exhibitions explain what they are about to the lay public. Progress in educating the public in general, and farmers in particular, is being made, but we still have a long way to go before the majority of farmers are involved to any extent in conservation.

There is still strong opposition from some farmers and their unions to measures which they think will restrict their freedom to manage their land in their own way. Thus in upper Teesdale in 1979 almost every farm sported a placard saying "No AONB", when it was proposed to designate that part of the country as an Area of Outstanding Natural Beauty. Most passing tourists had no idea what the initials AONB stood for, but the farmers found this ignorance an advantage, for when they explained their significance they were able to enlist sympathy against what they described as unwarranted interferance from city-based bureaucrats. Their campaign was successful, and the project was abandoned. Then in mid-Wales when the NCC decided to designate 21,449 hectares of uplands as a Site of Special Scientific Interest (SSSI), this was opposed by local farmers with the support of the NFU and the Farmers Union of Wales (one of the few occasions when these two bodies agreed).

There have been various proposals that the agricultural advisory services should also advise farmers on conservation. Some of the officers of ADAS are well qualified to do so, being themselves personally involved in various countryside problems. The trouble is that even the most expert authority may have some difficulty in knowing just what he should advise, in view of the fact that he will generally find that he is asking the farmer to make some sacrifice, even if it is a modest one, for which he will receive little financial return.

This brings us naturally to the vexed question of government grants. At the time of writing, in 1980, the government spends some £400 million each year in grants to farmers. These pay part of the costs of improving farm buildings, they aid waste disposal and drainage, they contribute to the costs of better water supply, they cover such activities as grubbing up orchards and some items of land improvement. Higher grants are paid to those farming in 'less-favoured areas' with Common Market assistance, and substantial sums or 'compensatory allowances' are available for breeding and rearing cattle in upland areas. Many conservationists compare the much smaller amount – less than £10 million in 1979 – paid to support the Nature Conservancy Council with the £400 million paid out in grants which may be wholly harmful to Britain's wildlife. They consider that, if farmers are to receive government grants, these should more often be used to compensate them for permitting wildlife to survive on their land.

Many prejudiced town dwellers look on these grants as part of the 'feather bedding' which farmers receive, and which are unfair to the rest of the tax payers who raise the money. The system is more complicated. British governments have had a policy of providing cheap food for the many urban voters, and have generally prevented farmers from receiving the full economic price for their produce. Grants are a way of putting more money into agriculture, and so raising at least some farmers' incomes, without directly raising food prices. Many farmers would prefer what they think would be a more honest system where higher prices provided all of their income. Under such a system most grants would not be necessary.

Until such a change is made I think that we should make sure that government money is not spent by one department in such a way that the work of another is destroyed. There has been some improvement in this direction though recent proposals to give retrospective approval of grants for work already completed are causing concern to conservation bodies such as the NCC and the SPNC. At one time when an application was made for a grant to drain an area of marshland, this could not be refused if it could be shown that the result would be to increase the agricultural productivity of the land, even if it was scheduled by the Nature Conservancy Council as a Grade 1* Site of Special Scientific Interest. Such a site is 'internationally important – the safeguarding of all Grade 1 sites is considered essential if there is to be an adequate basis for nature conservation in Britain, in terms of a balanced representation of ecosystems, and inclusion of the most important examples of wildlife or habitat'.

The government has made some concessions which should give a little help to conservation. In the past it has been obligatory for the planning authorities to consult the Nature Conservancy Council if there is any proposal to alter the use of a Site of Special Scientific Interest. Change of use has usually meant some type of non-agricultural development, housing or the erection of a factory. There is no guarantee that the advice of the NCC will be taken in such cases, and the planning authority has every right to agree to the destruction of the SSSI, but at least the conservation case is heard. Drainage of agricultural land was not considered to be a change of use as the land would continue to be farmed. Not only could the drainage not be prevented but, as has already been stated, a grant could not normally be refused. Things have now been changed so that grants for developments of SSSIs, even when these are purely agricultural, are not given automatically. The Nature Conservancy Council is consulted, and as a rule voluntary amenity groups have an opportunity of stating their views. In some cases applications have gone to public inquiry, and grants have not been approved.

In the past many owners of wetland SSSIs have drained them at their own expense, and have not bothered to apply for a grant, considering that the government machinery is too complicated and dilatory. If grants are now given for work already completed, when an SSSI or National Park is affected, the farmer is expected to make sure that the NCC or other planning authority approves of his plans. If they do not, the grant may be refused. However, from the point of view of conservation this is little consolation, as the damage to wildlife will already have been done.

It is high time that some change in this whole procedure was made. The Nature Conservancy Council has 165 National Nature Reserves, covering 131,851 ha, but much of this land is only safeguarded by short leases or agreements with landlords for fixed periods, many of which are already on the point of expiring. The NCC has also scheduled 3,750 SSSIs covering 1,160,000 ha, but these are being developed for other purposes at a rate of 4 per cent per annum, so many of the most important could soon be lost. Although over 500 SSSIs are now managed by voluntary bodies, the Society for the Promotion of Nature Conservation, the many County Naturalists' Trusts and the Royal Society for the Protection of Birds, the majority are still very vulnerable. We are only protecting the wildlife in a very small proportion of the total area of Britain, and much of the protection is far from permanent or complete.

Fortunately there are other ways in which some SSIs (and wildlife generally) may be safeguarded. There is various legislation, the National Parks and Access to the Countryside Act of 1949, as amended by the Nature Conservancy Council Act of 1973, the Countryside Act of 1968, etc. etc, which may help. A good summary of this legislation is included in *Nature Conservation and Agriculture*, the NCC pamphlet already mentioned.

One of the most important provisions comes under Section 15 of the Countryside Act 1968, which enables the Nature Conservancy Council to give financial support to owners who manage SSSIs in an agreed manner. Thus a farmer may have a meadow which has not been ploughed since the Black Death, rich with wild flowers. By reseeding and dressing liberally with fertilisers he could greatly increase its productivity, producing more hay or silage and grazing a greatly increased head of livestock, but, at the same time destroying the wild flowers. There are limited funds available to compensate the farmer for *not* improving his field. Similar arrangements may apply to marshes or old deciduous woodland. The difficulty is to determine a proper level of compensation. For many years the Treasury set this at such a low figure that farmers were not interested. Fortunately voluntary bodies, particularly County Naturalists' Trusts, using their own money raised from their members, were able to offer more realistic sums. We must all applaud the Treasury for trying to save the taxpayer's money, but this is not the first case where the poor judgment of our officials has had an effect opposite to that intended by the legislation. Now things are a little better, and payments similar to those agreed by voluntary bodies have been made, but the total funds are insufficient to protect more than a small fraction of the endangered sites.

There have also been provisions under various Finance Acts (e.g. 1975 and 1976) which may ensure that some National Nature Reserves remain as such. Thus if the management of land for a reserve is agreed with the Nature Conservancy Council, it is exempt from Capital Transfer Tax. Also if land is of outstanding conservation importance it may be transferred at death or as a gift to a new owner, if there is an undertaking to maintain the character of the land and to allow the public reasonable access to it. This may give more permanent value to Nature Reserve agreements under which many reserves are maintained.

Drainage grants have caused the most concern. It should be recognised, however, that there is no conservation objection to many of the projects so supported. The largest number are for underfield tile drainage of land which is already growing arable crops or artificially-seeded leys. Here the drainage is essential if the fields are to be effectively farmed, particularly in wet years. It is when natural or semi-natural wetland is drained that the damage is done to wildlife. As has already been indicated, some protection may be given to the most valuable wetlands which have been scheduled as SSSIs, but other bogs and marshes, which have nearly as much value, are rapidly disappearing. It is to be hoped that voluntary conservation bodies will quickly acquire as many as possible as local nature reserves.

One of the few grants which may aid conservation is that given to produce reservoirs to store water for irrigation. Many of these ponds are stark, plastic creations, but others offer considerable scope for wildlife. It should surely be possible that, where a grant is given, it should be only for reservoirs specially designed so that they can serve this dual purpose.

Every year many hectares of farmland are taken over for roads, airfields, housing and industrial purposes. There is some doubt as to the actual figure, but it is unlikely to be less than 12,000 ha and some authorities think it may be twice as much. The Ministry of Agriculture is, naturally, making a great effort to reduce this loss, and most planning departments are trying to cooperate to this end. This may mean that other land, of greater conservation importance, is used instead. Thus several County Councils are finding it difficult to obtain convenient sites for refuse disposal. Often the site which is in the best place is good agricultural land. Another area, perhaps of saltmarsh, may be fairly near. The Ministry will then exert pressure to preserve the farmland and tip the refuse on the marsh. Now while I agree that we must take every effort to stop the loss of agricultural land, I also think that we need a better system of priorities. The total area of saltmarsh of conservation value in Britain is very limited. Once it is used as a rubbish dump, it is destroyed for ever. If refuse is tipped on farmland this can, eventually, be restored to agriculture. Such restoration is not always complete, but in some cases the land is returned in better condition than before it was used for tipping. From a strategic point of view one reason for preserving the saltmarsh is that in an emergency it can still be drained and turned into farmland, much as this course would be regretted by conservationists.

So far I have deal with farming as it is practised today, and indicated some of the ways in which its harmful effects on wildlife may be mitigated. However, there may be further, far-reaching changes in farming practice in the future, and these could have their effects, harmful or beneficial, on wildlife. It is impossible to foretell the future with any certainty, but some of the possible alternative developments are worth discussing.

I shall start with the subject of pesticides. I have already indicated in chapter 9 that although there is public concern at their effects, these are today generally not as important as is often suggested, and also the situation is generally improving. With regard to herbicides, I expect this improvement to continue. Weedkillers are becoming more selective, better able to eradicate particular weeds. They are also becoming more expensive. There is no suggestion in Britain that many weeds are becoming resistant to herbicides. I expect to see better control of weeds within crops, and little damage to wild plants outside the cultivated area.

The position regarding insecticides and fungicides may not be so reassuring. Although most insect pests can today be successfully controlled by existing insecticides, we do have some which are already resistant to more than one of the most widely used chemicals. I agree with the Royal Commission on Environmental Pollution that there is a grave risk of this process of resistance becoming more common, so that more and more insecticidal chemicals will cease to be capable of protecting our crops and livestock. It is only sensible that we should make every effort to prolong the useful life of the chemicals we now have, as we have no certainty that new replacements will be prepared in the near future. This means that every effort should be made to reduce pesticide use except when this is really necessary. This can only be good for wildlife, as accidental poisoning will obviously be reduced. We must avoid the opposite reaction, that is the use of greater and greater doses of chemicals in an attempt to kill resistant pests. This would certainly endanger non-pest species.

In chapter 9 I expressed some doubt as to whether biological control, that is the use of natural enemies of pests, would ever be very effective in annual crops. Not everyone agrees with this point of view, and hope is expressed that at least some contribution to pest control will be made, particularly if crops are grown in smaller plots, with greater diversity within quite restricted areas, and possibly with the encouragement of wild plants, even of 'weeds', which may increase the number of parasites and predators. While few scientists believe these measures would be completely effective, some think that they could contribute to 'integrated control' with minimal use of chemicals. This might make a minor contribution to wildlife survival. In a diverse situation more wild plants would survive, as would the animals dependent on them.

One way of reducing the use of chemicals is to revert to the cultural methods which were practised before the 1939 war when few effective insecticides were available. Unfortunately this might mean reducing our crop yields to the levels which applied at that time. But by adjusting planting dates and crop rotations quite reasonable yields might be obtained. Organic farmers will no doubt continue to try to establish whether their methods can be relied on to reduce pest damage sufficiently to make the use of toxic chemicals unnecessary. Plant breeders will continue to try to produce cultivars which are distasteful to pest insects and unattractive to rusts, mildews and other fungi.

There is even more serious concern that fungal diseases may become increasingly uncontrollable. Resistant varieties of crop plants seem to be attacked after a few years, as new strains of parasitic fungi appear. Fungicides seem effective only for limited periods. Here mixed cropping may prolong the useful life of fungicides, and may delay the appearance of new fungicide-resistant strains of fungi.

It must be said at this point that by no means all agricultural scientists are as concerned about resistance and the possibility that pesticides may become increasingly ineffective. They, and most farmers, take an optimistic point of view, and assume that new and more effective chemicals will continue to be available. This is clearly important to the farmer, but I do not believe that the survival of wildlife will be greatly affected whether or not crop pests are effectively controlled by chemicals or by other means. I believe that our control measures, including the Pesticide Safety Precautions Scheme, will make it unlikely that new chemicals with serious and undesirable ecological effects will be widely used. Improved techniques for spraying, using smaller amounts of chemicals but making sure that they reach their targets, will probably reduce the danger to other organisms including wildlife.

The future pattern of farming in Britain depends more on economic and political forces than on the biology of pests and diseases. The government suggests that we should continue to try to increase productivity and become more self-sufficient. This policy may be affected by the ways in which our partners in the Common Market progress;

where there is a surplus of a particular commodity, Britain may be encouraged to reduce her efforts to produce it. Many people consider that the growing world population will reduce the amount of food available for import, and will increase its price, so we will be compelled to become more self-sufficient. Others think that there may be global overproduction for many years to come. The energy crisis, with dearer and scarcer fuel, may have its effects on our present energy-intensive agricultural system. I am here concerned with the ways in which wildlife may be affected by all these possibilities.

The official view is that recent trends in British farming will continue. Thus in the report of the Royal Commission on Environmental Pollution in an appendix prepared by the Ministry of Agriculture, Fisheries and Food, we read: '. . . the Ministry foresees further improvement in efficiency, with higher yields and stocking densities, and a substantial increase in output from bigger and more highly mechanized farms using less land overall, with fewer separate activities on each, and using a smaller labour force. There will be less, but more intensively managed, grassland; more livestock will be in big units; but there will be little change in the current pattern and location of crops. In arable farming, trends to minimum cultivation are expected to continue. The use of fertilisers and pesticides will have increasing importance; but with more precise application techniques and other improvements, there will be only a moderate overall increase in the total volume of chemicals used. In dairying, the Ministry expects bigger herds, improved yields, greater use of fertilisers and a continuing trend to cubicle housing. No further dramatic changes are expected in the overall pattern of poultry production; pig units will continue to grow in size; and efficiency in both pigs and poultry will improve further. Stocking densities of beef cattle and sheep will rise'.

This forecast suggests greater intensity on the better land, but a reduction of the area of Britain which will be farmed. This will partly compensate for the loss of farmland to building, roads, etc., but it could make more available for conservation. It is indeed possible that the cultivated area of Britain may decrease substantially. This has happened over the last two hundred years in Europe, and in this century many of the farms in New England in the United States have gone out of cultivation and have reverted to 'wilderness'. As productivity per hectare has increased the arable area has decreased, notwithstanding the need to feed an increasing population. If this scenario is the right one, the argument that we need to drain more marshes and improve old grass ceases to be tenable, so the possibility of increasing the area devoted to conservation improves.

This optimistic forecast for both agriculture and conservation may not be realised for various reasons. First, our climate may change. Some meteorologists believe that we are entering another ice age, and though it may be many hundreds or even thousands of years before Britain is again largely covered with ice, there might be a fall of one degree centigrade in our mean temperature by the year 2,000. Others suggest that the increased production of carbon dioxide from burning coal, oil and natural gas may produce a 'greenhouse effect' with a rise in temperature of the globe. These temperature changes would also affect the rainfall pattern, parts of the world which are now adequately watered might become dry, and deserts might have a substantial rainfall. Climatic changes would affect both farming and wildlife. Even a very small fall in mean temperature might reduce the growing season for both grass and cereals by some weeks. Our wildlife would be equally affected, for instance by driving arctic species of bird south. I find that the most alarmist views on possible climatic change are voiced by those with little knowledge of this subject, and that the scientists who have devoted most time to this study take a more conservative point of view. I think that we should monitor all possible climatic changes, we should make outline plans when there is evidence that changes are taking place, but that today neither farming nor conservation policy should be affected by what are purely hypothetical changes which may not take place.

The world energy crisis will have its effects on farming. Since the war we have replaced horses with tractors, and have reduced the manpower engaged in agriculture by two-thirds. There are those who think this process is likely to be reversed. I do not agree. We spend less than 4 per cent of our total national energy budget on our farms; surely we will continue by some system of priorities to give this support to food production. However, high prices of fuels have already had some effect. Direct drilling saves nearly three-quarters of the fuel used in ploughing and cultivating the land. Herbicides, by reducing the amount of cultivation, save a great deal of energy. Horses are unlikely to make a significant comeback, partly because they use too much energy, partly because they are too slow. The energy in the food eaten by horses is far greater than that in the oil used by tractors doing the same work. We could not harvest the 17 million tonnes of cereals grown in this country with slow, horse-drawn equipment in the time available. Also arable farming is very energy-efficient. The grain harvested from winter wheat contains at least three times the energy used to grow the crop, allowing for the energy used to make the machines, to operate them, to produce the fertiliser and to harvest the crop. The straw itself could supply enough energy to run all our agricultural machinery, if it could be transformed into a more usable form. So I foresee little change in arable farming, except that high costs will mean that methods which save fuel will become more popular.

We actually burn more oil to heat greenhouses than we use to run farm machinery. This is surely one place in which economies are likely to be made, as the flowers and out-of-season vegetables produced cannot be considered essential to our survival. I expect that waste heat from power stations will be more widely used to heat glasshouses, which may in the future be found concentrated in places like the Trent valley where much of our electricity is generated. Perhaps every nuclear power station will also support many acres of glass.

The greatest part of the energy used in agriculture in Britain is to produce nitrogenous fertilisers. The rise in oil prices has meant that these chemicals are expensive, and so farmers wish to economise in their use and still obtain the high yields they have generated. When chemicals were cheap, they were preferred to farm yard manures which were more troublesome, and required more labour, to apply. Today there is more economic pressure to use natural manures, even to transport them from livestock units to all-arable farms. We may see a return to mixed farms (though the Ministry of Agriculture does not expect this to happen) so that manure is produced where it can be used. Organic farming should become increasingly popular. These changes may give rise to a modest increase in the number of farm workers. They will probably also be beneficial to wildlife.

It may perhaps be worth mentioning that the food processing industry uses more than twice as much energy as is used on the farms of Britain. This figure is sometimes added to that used by the agricultural industry itself, causing people to get the wrong idea about the industry's efficiency. While there is some room for more economy in energy use on the farm, there is clearly much more scope for saving in processing, packaging and transporting it to the customer.

There are those who advocate drastic changes in our whole farming system. They would like to see our large enterprises broken up to many smaller units, employing a far greater proportion of our population. Such changes might possibly be beneficial to the whole rural economy, they could make their contribution to a reduction in unemployment and wildlife would be considerably affected. I do not think that there will be a widespread revolution on these lines, though there is a growing number of people wishing to return to a life of rural self-sufficiency. It is often said that many smaller farms would be more efficient in producing food than is the present system. This statement is based on a misunderstanding of the statistics. Small farms and small-holdings may grow expensive and labour-intensive crops, so producing a substantial cash income per unit area. Intensive livestock units on restricted sites appear very efficient, until it is realised that the animals live on food produced on large farms elsewhere. In fact the highest yields of crops like cereals, sugar beet and potatoes are generally found on the largest and most highly mechanised farms, and the most highly yielding cattle live on the large dairy farms. Also small, intensive farms have little room for wildlife.

There is probably little future for the small hill farm. We see heart-rending programmes on television, showing the economic problems of the man who is trying to make a living from a few ancestral hectares of poor hill land. It is clearly impossible to make an income comparable to that of a factory worker from a score of sheep and a few bullocks which is all the land will support. What may not be realised is that this sort of farm does not provide its owner with a day's work either. Of course Parkinson's law – 'Work expands to fill the time available' – applies to farms; the hill farmer can keep himself busy tending his few animals and making a little hay for their winter keep. But he could equally well have looked after a hundred times as many animals, as does the shepherd on the nearby moor. I see no future for the small upland farm as a full-time occupation, and believe that further amalgamation is inevitable. Unfortunately the larger area will be better managed, which means that the grass will be improved and its conservation interest reduced.

In conclusion, I generally agree with the Ministry that it is most probable that farming in the future will continue much on its present lines. This means that the scope for wildlife conservation on farmland is limited. I hope that all the types of compromise which we identify with 'Silsoe' will continue and be extended to more and more farms. Only thus will at least some of our wildlife continue to exist in most of rural Britain. But the need for more nature reserves, managed specifi-cally for wildlife, is increasingly obvious. Only thus can many ecosys-tems and rare species be preserved. I would like to see at least some reserves set up and carefully managed on good, Grade 1 or Grade 2, farmland, where wildlife would flourish as well as do arable crops. At present only a negligible area of our reserves is on anything but the most unproductive Grade 5 land. I believe that we in this country must be prepared to make this sort of sacrifice; the Ministry's statement suggests that the land should be available. We must also make far greater efforts to protect the decreasing area of wetlands, old grass and other habitats if we wish to protect and preserve our wildlife so that it may be enjoyed by future generations. These areas are not, at the moment, really essential to agriculture, and once they are 'improved' they will be permanently destroyed as habitats for our unique native flora and fauna.

POSTSCRIPT

The following article is one which I wrote in 1975 and which appeared in the weekly scientific journal *Nature*.

Theology was once known as 'the queen of the sciences', and although an examination of recent *Nature* indexes does not produce many references to this subject, I think it may be permissible to mention the idea of the 'God of the gaps'. I have unfortunately been unable to trace the origin of this phrase, which is used to describe the views of those who are unable to come out clearly as rationalistic atheists. They hang on to the idea of a deity, but restrict his activities to the gaps between the major fields of man's activities, where science is thought to be able to give a full explanation.

My interest in this theological proposition was aroused when I realised that it was very similar to the view held by many leading conservationists. They say that they believe that wildlife and country-side preservation is important, but they acknowledge that the interests of the farmer and of food production must have priority in any scheme for managing the rural landscape. This point is explicit in the recent report of the Countryside Commission, *New Agricultural Land-scapes*. Any wish to retain the familiar pattern of hedgerows, pictur-esque buildings and flower-rich though rather unproductive meadows is castigated as sentimental. We are urged to try to make the best of the inevitable.

It must be admitted that, even within this pattern of farming development, useful compromises which have done much to spare wildlife (particularly birds) have been possible. There have now been many 'Silsoe-type' exercises, for example. These follow the pattern of the first weekend conference at Silsoe in Bedfordshire, when agricul-turalists and conservationists tried to work out various schemes for managing a farm where nearly maximum productivity could be married with the least damage to the native flora and fauna. Small, unproductive patches on the farm are identified, and these are planted with trees and shrubs to act as mini-nature reserves.

Notwithstanding these developments, however, some impoverishment of the landscape and a reduction in numbers of many native plants and animals is inevitable. This is accepted because of the belief that we must do everything possible to maximise food production, as otherwise we may all face malnutrition at the best and mass starvation at the worst. A country like Britain that imports nearly half its food cannot enjoy the luxury of 'wasting' any area for conservation if it can be used for food production.

I believe that the time has come for conservationists to be much more aggressive. Farmers have a right to make a reasonably good living from what may be a very difficult and exhausting job. The government has the duty to see that Britain's food supply is safeguarded, even if the pound sterling falls in value so that we cannot continue to import so much of what we eat. There is a good chance that in a few years the growing world population will absorb any surpluses, and that we will be unable to make good food deficiencies by imports. All these points are taken as arguments in favour of maximising productivity and treating conservation with caution.

I do not believe, however, that the choice is really between the risk of starvation with a rich and varied countryside, and enough food with an impoverished landscape. Britain already produces enough food to provide its population with an adequate diet, with enough calories and protein for all. Imports are mainly used to feed livestock to provide meat, which is produced for enjoyment rather than to prevent any malnutrition. The choice is really between two forms of enjoyment – a meat-rich diet or a countryside rich in wildlife. So the conservationist need no longer be content with the gaps – he can come out in to the open and rightly demand his share in shaping the future pattern of all parts of the rural landscape.

BIBLIOGRAPHY

Books and papers relevant to the different topics discussed in this book are listed under the appropriate chapter headings. However, many of the most important references deal with several of the subjects, and with the general problems of wildlife conservation and agriculture. These more general references are included under Chapter 12.

CHAPTER 1 *Introduction*

CLAPHAM, A. R., TUTIN, T. G. and WARBURG, E. F. (1962). *Flora of the British Isles*, 2nd Edition. London, Cambridge University Press.

CORBET, G. B. and SOUTHERN, H. N. (1977). *The Handbook of British Mammals*, 2nd Edition. Oxford, Blackwell Scientific Publications.

HAWKSWORTH, D. L. (Ed.) (1974). *The Changing Flora and Fauna of Britain*. London, Academic Press.

KLOET, G. S. and HINCKS, W. D. (1945). *A Check List of British Insects*. Arbroath, Buncle.

LEVER, C. (1977). *The Naturalized Animals of the British Isles*. London, Hutchinson.

MATTHEWS, L. H. (1972). *British Mammals*. London, Collins New Naturalist.

MINISTRY OF AGRICULTURE, FISHERIES AND FOOD (1975). *Food From Our Own Resources*. London, HMSO.

MURTON, R. K. (1971). *Man and Birds*. London, Collins New Naturalist.

PARSLOW, J. (1973). *Breeding Birds of Britain and Ireland*. Berkhamsted, Poyser.

PERRING, F. (Ed.) (1970). *The Flora of a Changing Britain*. Hampton, Classey, for the Botanical Society of the British Isles.

SALISBURY, E. (1964). *Weeds and Aliens*. London, Collins New Naturalist.

STAMP, L. D. (1969). *Man and the Land*. London, Collins New Naturalist.

TAYLOR, J. C. (1979). The introduction of exotic plant and animal species into Britain. *Biologist*, 26, pp.229–36.

WORMELL, P. (1978). *Anatomy of Agriculture, A Study of Britain's Greatest Industry*. London, Harrap and Kluwer.

CHAPTER 3 *Grass and Grazing*

CLAPHAM, A. R. (Ed.) (1978). *Upper Teesdale. The Area and Its Natural History*. London, Collins.

DEPARTMENT OF AGRICULTURE AND FISHERIES FOR SCOTLAND and THE NATURE CONSERVANCY COUNCIL (1977). *A Guide to Good Muirburn Practice*. London, HMSO.

DUFFEY, E., MORRIS, M. G., SHEAIL, J., WARD, L. K., WELLS, D.A., and WELLS, T. C. E. (1974). *Grassland Ecology and Wildlife Management*. London, Chapman and Hall.

GIMINGHAM, C. H. (1972). *Ecology of Heathlands*. London, Chapman and Hall.

MOORE, I. (1966). *Grass and Grassland*. London, Collins New Naturalist.

WELLS, T. C. E. (Ed.) (1965). Grazing Experiments and the Use of Grazing as a Conservation Tool. *Monks Wood Experimental Station Symposium No. 2*.

WILLIAMS, O. B., WELLS, T. C. E. and WELLS, D. A. (1974). Grazing management of Woodwalton Fen: Seasonal changes in the diet of cattle and rabbits. *Journal of Applied Ecology*, 11, pp.499–516.

CHAPTER 5 *The Living Soil*

BALFOUR, E. B. (1975). *The Living Soil and the Haughley Experiment*. London, Faber and Faber.

EDWARDS, C. A. and LOFTY, J. R. (1972). *Biology of Earthworms*. London, Chapman and Hall.

GRIFFIN, D. M. (1972). *Ecology of Soil Fungi*. London, Chapman and Hall.

KEVAN, D. K. McE. (Ed.) (1955). *Soil Zoology*. London, Butterworth.

KUHNELT, W. (1961). *Soil Biology, with Special Reference to the Animal Kingdom*. London, Faber and Faber.

MELLANBY, K. (1971). *The Mole*. London, Collins New Naturalist.

RUSSELL, E. J. (1957). *The World of the Soil*. London, Collins New Naturalist.

CHAPTER 6 *Hedges and Trees*

CENTRE FOR AGRICULTURAL STRATEGY (1980). *Report No. 6. Strategy for the UK Forest Industry*. Reading.

EDLIN, H. L. (1956). *Trees, Woods and Man*. London, Collins New Naturalist.

HOSKYN, C. W. (1857). *Talpa or the Chronicles of a Clay Farm*. London, Longman, Brown, Green, Longmans and Roberts.

McNAUGHT, K. (1980). The New Hayley Lane Hedge. *Nature in Cambridgeshire*, 23, pp.26–7.

POLLARD, E., HOOPER, M. D. and MOORE, N. W. (1974). *Hedges*. London, Collins New Naturalist.

Wells, T. C. E. (1980). Management options for lowland grassland. From Rorison, I. H. and Hunt, R. (1980) Amenity Grassland: An Ecological Perspective. Chichester, John Wiley.

CHAPTER 7   *Ponds, Rivers and Marshes*

Astbury, A. K. (1958). *The Black Fens*. Cambridge, Golden Head Press.
Dyson, J. (1976). *The Pond Book*. London, Kestrel.
Macan, T. T. and Worthington, E. B. (1951). *Life in Lakes and Rivers*. London, Collins New Naturalist.
Owens, M., Garland, J. H. N., Hart, I. C. and Wood, G. (1972). Nutrient budgets in rivers. *Symposium of the Zoological Society of London*, No. 29, pp.21–40.
Ranwell, D. S. (1972). *Ecology of Salt Marshes and Sand Dunes*. London, Chapman and Hall.
Wentworth-day, J. (1954). *A History of the Fens*. London, G. Harrap.

CHAPTER 8   *Farm Houses, Yards, Gardens and Barns*

Campbell, B. (1974). Birds of an Oxfordshire oasis. *Country Life*, 155, pp.774–5.
Harvey, N. (1970). *A History of Farm Buildings in England and Wales*. Newton Abbot, David and Charles.
Hickin, N. E. (1964). *Household Insect Pests*. London, Hutchinson.
Mourier, H. and Winding, O. (1975). *Collins Guide to Wild Life in House and Home*. London, Collins.
Owen, D. (1977). Hoverflies in the garden. *Country Life*, 161, pp.658–63.

CHAPTER 9   *Pests and Pesticides*

Edwards, C. A. and Heath, G. W. (1964). *The Principles of Agricultural Entomology*. London, Chapman and Hall.
Gunn, D. L. and Stevens, J. G. R. (Eds.) (1976). *Pesticides and Human Welfare*. Oxford, University Press.
Irvine, D. E. G. and Knights, B. (Eds.) (1974). *Pollution and the Use of Chemicals in Agriculture*. London, Butterworth.
Jones, F. G. W. and Jones, M. G. (1964). *Pests of Field Crops*. London, Arnold.
Mellanby, K. (1967). *Pesticides and Pollution*. London, Collins New Naturalist.
Mellanby, K. (1980). *The Biology of Pollution*, 2nd Edition. *Studies in Biology, The Institute of Biology*. London, Arnold.
Ministry of Agriculture, Fisheries and Food (1980). *Approved Products for Farmers and Growers*. London, HMSO.

PERRING, F. H. and MELLANBY, K. (Eds.) (1977). *Ecological Effects of Pesticides. Linnean Society Symposium Series.* London, Academic Press.
ROYAL COMMISSION ON ENVIRONMENTAL POLLUTION (1979). *Seventh Report. Agriculture and Pollution.* Cmnd 7644. London, HMSO.
STAPLEY, J. H. (1949). *Pests of Farm Crops.* London, Spon.

CHAPTER 10   *Field Sports*

THE GAME CONSERVANCY (1980). *Game on the Farm.* Fordingbridge.
PAGE, R. (1977). *The Hunter and the Hunted, A Countryman's View of Blood Sports.* London, Davis-Poynter.
VESEY-FITZGERALD, B. (1946). *British Game.* London, Collins New Naturalist.

CHAPTER 11   *Zoonoses and Diseases of Livestock Involving Wild Mammals and Birds*

ANDREWES, C. H. and WALTON, J. R. (1977). *Viral and Bacterial Zoonoses.* London, Bailliere Tindall.
FIENNES, R. (R. N. TWISLETON-WYKEHAM-FIENNES) (1978). *Zoonoses and the Origins and Ecology of Human Disease.* London, New York, San Francisco, Academic Press.
GRAHAM-JONES, O. (Ed.) (1968). *Some Diseases of Animals Communicable to Man in Britain. Proceedings of a Symposium Organised by the British Veterinary Association and the British Small Animal Veterinary Association, London, June 1966.* London, Pergamon.
HULL, T. G. (Ed.) (1963). *Diseases Transmitted from Animals to Man.* Springfield, Ill., Thomas.
McDAIRMID, A. (Ed.) (1969). *Diseases in Free-Living Wild Animals.* London, Zoological Society and Academic Press.
MINISTRY OF AGRICULTURE, FISHERIES AND FOOD (1979). *Bovine Tuberculosis in Badgers, 3rd Report.* London, HMSO.
MINISTRY OF AGRICULTURE, FISHERIES AND FOOD (1980). Badgers, cattle and tuberculosis, by Lord Zuckerman.
MURTON, R. K. (1964). Do birds transmit foot and mouth disease? *Ibis,* 106, pp.289–98.
STODDART, D. M. (Ed.) (1979). *Ecology of Small Mammals.* London, Chapman and Hall.

CHAPTER 12   *Wildlife Conservation on the Farm*

AGRICULTURAL DEVELOPMENT AND ADVISORY SERVICE (1972). *Churn Farming and Conservation Study: The Report of an Exercise and Conference. . . .* Churn Estate, Blewbury, Berkshire. London, HMSO.

AGRICULTURAL DEVELOPMENT AND ADVISORY SERVICE (1973). *Agriculture in the Urban Fringe: A Survey of the Slough/Hillingdon Area*. London, HMSO (Technical Report 30).

AGRICULTURAL DEVELOPMENT AND ADVISORY SERVICE (1974). *The Cowbyers Conference on Upland Farming, Forestry, Game Conservation and Wildlife Conservation, 12, 13, 14 July 1974: [organised] with the Countryside Commission*. Newcastle upon Tyne.

AGRICULTURAL DEVELOPMENT AND ADVISORY SERVICE (and others) (1976). *Farming Wildlife and Landscape: Essex Exercise, June 1975*. Colchester, Essex County Council.

AGRICULTURAL DEVELOPMENT AND ADVISORY SERVICE (1976). *Wildlife Conservation in semi-natural Habitats on Farms; A Survey of Farmer Attitudes and Intentions in England and Wales*. London, HMSO.

ADVISORY COUNCIL FOR AGRICULTURE AND HORTICULTURE IN ENGLAND AND WALES, NIGEL STRUTT, CHAIRMAN (1978). *Agriculture and the Countryside*. HMSO.

BARBER, D. (Ed.) (1970). *Farming and Wildlife: A Study in Compromise*. Sandy, Royal Society for the Protection of Birds.

CARTER, E. S. and SAYCE, R. B. (1979). Conservation and agricultural land. Journal of the Royal Agricultural Society of England, Vol. 140, pp. 22–33

CHESHIRE COLLEGE OF AGRICULTURE (1976). *Wildlife and Landscape Management Plan*, ed. by S. E. Crooks. Chester, Cheshire County Council Education Committee.

COUNCIL FOR THE PROTECTION OF RURAL ENGLAND (1971). *Loss of cover through the removal of hedgerows and trees. Report of a working party*.

COUNTRYSIDE COMMISSION (1974). *New Agricultural Landscapes*. London, HMSO.

COUNTRYSIDE COMMISSION (1978). *Upland Land Use in England and Wales*. London, HMSO.

DAVIDSON, J. and LLOYD, R. (Eds.) (1977). *Conservation and Agriculture*. Chichester, John Wiley.

ENERGY AND AGRICULTURE, Special Issue. *Span*, 18 (1), (1975).

FARMING AND WILDLIFE ADVISORY GROUP (1973). *The Chalkland Exercise on Chalkland Farming and Wildlife Conservation, 20, 21 and 22 July 1973, Kingston Deverill, Wiltshire*. College of Sarum St. Michael, Salisbury; London, Agricultural Development and Advisory Service.

FARMING AND WILDLIFE ADVISORY GROUP (1974). *Farming and the Countryside Exercise for Suffolk 1974*. Sandy.

FARMING AND WILDLIFE ADVISORY GROUP (1975). *Report of the Dinas Conference on Upland Farming, Forestry and Wildlife Conservation, 11, 12, 13 August 1972*. St. David's, University College, Lampeter. 2nd edition, Pinner.

HARVEY, G. (1980). Profits and wildlife can go hand in hand. *Farmers Weekly*, 92 (17), xviii–xxi.

HAWKES, J. G. (Ed.) (1978). *Conservation and Agriculture*. London, Duckworth.

LEACH, G. (1976). *Energy and Food Production*. Guildford, IPC Press.

LEICESTERSHIRE RURAL COMMUNITY COUNCIL (1976). *Prestwold – 2000: A Report on the Future of a Large Estate*. Leicester.

MABEY, R. (1980). *The Common Ground: A Place for Nature in Britain's Future*. London, Hutchinson.

MELLANBY, K. (1975). *Can Britain Feed Itself?* London, Merlin.

MINISTRY OF AGRICULTURE, FISHERIES AND FOOD (1979). *At the Farmer's Service, A Handy Reference to Various Services Available to Farmers in England and Wales*. London, HMSO.

MINISTRY OF AGRICULTURE, FISHERIES AND FOOD (1979). *Farming and the Nation*. Cmnd 7458. London, HMSO.

MINISTRY OF AGRICULTURE, FISHERIES AND FOOD (1979). *Possible Patterns of Agricultural Production in the United Kingdom by 1983*. London, HMSO.

NATURE CONSERVANCY COUNCIL (1977). *Nature Conservation and Agriculture*. London.

Neighbours with nature: A look at farming's relationships with the wildlife in woodland, wetland and wasteland. *Farmers Weekly*, 31 December 1976, pp.49–62.

PIMENTAL, D. and PIMENTAL, M. (1979). *Food, Energy and Society*. London, Arnold.

RITSON, C. (1980). *Self-sufficiency and Food Security*. Centre for Agricultural Strategy, Paper 8. Reading.

RUCK, A. (1980). *Nature Conservation. Why and How? – An Introduction*. Lincoln, SPNC.

SHOARD, M. (1980). *The Theft of the Countryside*. London, Temple Smith.

STUDY CONFERENCE ON FARMING AND WILDLIFE IN DORSET [1971]. *Farming and wildlife in Dorset: the report of an exercise and conference which took place at East Farm, Hammoon, and at Weymouth College of Education, 25 and 26 July 1970*. Weymouth College of Education.

SWANNICK, C. (Ed.) (1979). *Eysey Farm, Farming and Wildlife Study*. Royal Agricultural College, Cirencester.

INDEX

When they are in general use, only English common names are given for plants and animals in this Index. Latin names will be found in the text on the first occasion when an organism is mentioned.

175